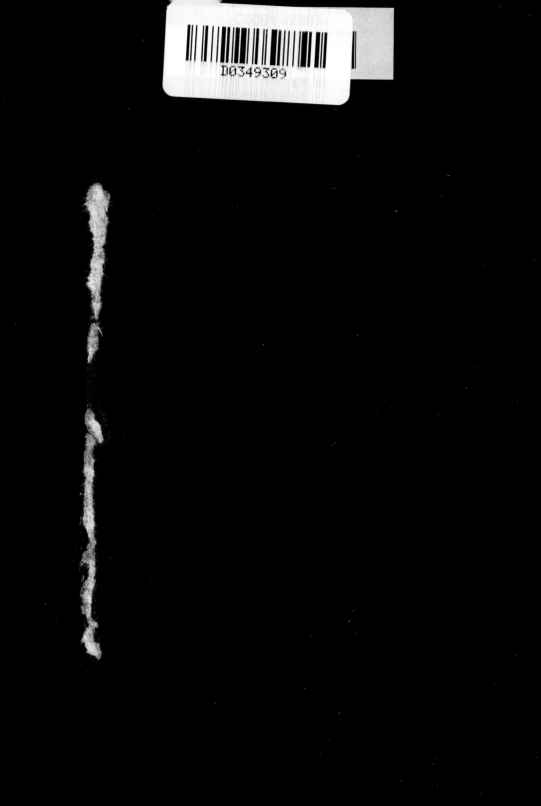

TOUCH

TOUCH

The Science of Hand, Heart, and Mind

DAVID J. LINDEN

VIKING

an imprint of

PENGUIN BOOKS

VIKING

UK | USA | Canada | Ireland | Australia
India | New Zealand | South Africa

Penguin Books is part of the Penguin Random House group of companies
whose addresses can be found at global.penguinrandomhouse.com.

First published in the United States of America by Viking Penguin,
a member of Penguin Group (USA) LLC 2015
Published in Great Britain by Viking 2015
001

Printed in Great Britain by Clays Ltd, St Ives plc

A CIP catalogue record for this book is available from the British Library

HARDBACK ISBN: 978–0–241–18403–5
TRADE PAPERBACK ISBN: 978–0–241–18404–2

In memory of Professor Steven Hsiao
A brilliant touch researcher
A warm friend
A man of hand, heart, and mind

CONTENTS

If soul may look and body touch, Which is the more blest?

—W. B. Yeats, "The Lady's Second Song," 1938

Seeing's believing, but feeling's the truth.

—Thomas Fuller, *Gnomologia*, 1732

PROLOGUE

Malibu, Summer 1975

We're eight teenage campers huddled around a fire ring, late at night. Piled up like puppies, spilling over rocks and stumps and the dusty bare dirt of the Santa Monica mountains, we smell of black sage and acorns and unwashed T-shirts. With no adults in sight and the soft cover of darkness, we give voice to our innermost pubescent thoughts.

"Your turn, Sam."

"Okay . . . this is for Caroline. Would you rather give an open-mouth kiss to the camp director or eat a live cockroach?"

Our voices rise in a disgusted, delighted Greek chorus, "Eeeeeeeeeeeew!"

"You're so gross, Sam. I'm not answering that one."

"But you have to. Those are the rules."

"No way, you pervert."

"You're so prickly. I didn't mean to hurt your feelings."

"Yeah, right."

"Okay, here's a clean one. Would you rather die of cold in Antarctica or heat in the Sahara Desert?"

"I'm not allowed to bring a parka to Antarctica?"

"No, you're naked."

"Then I choose the desert. I want to go out with a good tan."

Good-natured howling erupts. Caroline raises her arm and shimmies, vamping it up.

Sam smiles. "You're so vain. And . . . I've gotta go." Everyone knows that this is bogus. It's obvious that he's crazy about Caroline.

"No, you don't, you slippery sonofabitch. Now it's *my* turn. You must give up all of your senses except one. Which do you pick to save?"

"Oh, man. That's rough. I'd keep sight. Then, at least I could get around. Uh, no, hearing—I need my music. Shit. I dunno. That would just suck."

"Yeah, it would."

"I'm touched by your concern."

"Bite me."

Later, lying in my sleeping bag and mulling over this flirtatious banter, I was puzzled. Flush with hormones, we all hungered for interpersonal touch, for kisses and caresses and more. I was typical of this group, so consumed with the idea of holding and kissing a lovely dark-haired girl named Lorelei that I could barely speak. Touch was central to our obsessions and fantasies, yet none of us ever chose to preserve it when, in the nights that followed, Caroline's question about losing one of our senses returned in the "Would you rather . . . ?" game. Did we simply not think the ramifications through? It's certainly true that a bunch of horny sleep-deprived amped-up teenagers sitting around a campfire is not the ideal forum for contemplation. Or was it that we could easily imagine what it would be like to experience the loss of sight or hearing (we had all shut our eyes or plugged our ears), or even of taste or smell, yet none of us had ever actually been able to re-create the sensation of the loss of touch. Perhaps touch was woven so deeply into our sense of self that we could not truly imagine life without it. Years later when I read *Lolita*, I discovered that Vladimir

Nabokov had, as usual, raised this very issue many years before: "It is strange that the tactile sense, which is so infinitely less precious to men than sight, becomes at critical moments our main, if not only, handle to reality."

For Nabokov's Humbert Humbert, touch was so infinitely precious an experience that even the merest tactile contact with his beloved Lolita aroused overwhelming passions. For all of us, the experience of touch is intrinsically emotional, and this is reflected in common expressions in English. Read the dialogue that opens this chapter and notice that phrases like "I'm *touched* by your concern" or "I didn't mean to hurt your *feelings*" and texture metaphors like "you're so *prickly*" or "that's *rough*" or "you *slippery* sonofabitch" didn't stand out at all. We are completely accustomed to describing a wide range of human emotions, actions, and personalities in terms of our skin senses:

"I was touched by her thoughtfulness."

"It's a sticky situation."

"That's enough of that coarse language."

"That is one hairy problem."

"He rubs me the wrong way."

In everyday speech, the tactile is so entangled with the emotional that when we encounter someone who is emotionally clumsy, we call him *tactless*: Literally, he lacks touch.

This may seem like a silly question, but it's not: Why are emotions called *feelings* and not *sightings* or *smellings*? Do touch metaphors really tell us something about the skin senses and their relationship to human cognition, or are they merely a common usage of present-day English? In fact, the constructions "I'm touched" to mean "I'm emotionally affected" and "my feelings" to mean "my tender emotions" have been in use in the language since at least the late thirteenth century. And such expressions are not unique to English, or even the Indo-European language group, as they are found in tongues as diverse as Basque and Chinese.

~~~~~~

People who are blind or deaf from birth will for the most part develop normal bodies and brains (apart from the visual or auditory areas) and can live rich and fruitful lives. But deprive a newborn of social touch, as occurred in grossly understaffed Romanian orphanages in the 1980s and 1990s, and a disaster unfolds: Growth is slowed, compulsive rocking and other self-soothing behaviors emerge, and, if not rectified, emergent disorders of mood, cognition, and self-control can persist through adulthood. Fortunately, even a relatively minor intervention—an hour per day of touch and limb manipulation from a caregiver—can reverse this terrible course if applied early in life. Touch is not optional for human development. We have the longest childhoods of any animal—there is no other creature whose five-year-old offspring cannot live independently. If our long childhoods are not filled with touch, particularly loving, interpersonal touch, the consequences are dramatic.

The critical role of touch in early development has not always been appreciated. Child-rearing advice of the 1920s from the psychologist John B. Watson (the founder of the psychological movement called behaviorism) cautioned parents about spoiling their children with physical affection: "Let your behavior always be objective and kindly firm. Never hug and kiss them. Never let them sit on your lap. If you must, kiss them once on the forehead when they say goodnight. Shake hands with them in the morning. Give them a pat on the head if they have made an extraordinarily good job of a difficult task."[1]

While most parents today do not restrict contact with their children to an occasional pat on the head, it's a different story outside the family. In our zeal to protect kids from sexual predators, we have promoted no-touch policies for teachers, coaches, and other supervisory adults that, while well meaning, have the inadvertent effect of adding to the touch deprivation of our children. As these kids have grown up in a touch-phobic environment and propagated these fears to their own children, our society as a whole has become further impoverished.

You may ask, "Okay, I understand that kids are sensitive, but once we've become adults, why does it matter if we're touch-deprived? This touchy-feely stuff is for hippies and time wasters. Just squirt out another glob of hand sanitizer (with that deeply satisfying *blorp* sound) and get back to work." The answer is that interpersonal touch is a crucial form of social glue. It can bind sexual partners into lasting couples. It reinforces bonds between parents and their children and between siblings. It connects people in the community and in the workplace, fostering emotions of gratitude, sympathy, and trust. People who are gently touched by a server in a restaurant tend to leave larger tips. Doctors who touch their patients are rated as more caring, and their patients have reduced stress-hormone levels and better medical outcomes. Even people with clipboards at the mall are more likely to get you to sign their petitions or take their surveys if they touch your arm lightly.

~~~~~~

The main point of this book is not merely to argue that touch is good or even that touch is important. Rather, it's to explain that the particular organization of our body's touch circuits, from skin to nerves to brain, is a weird, complex, and often counterintuitive system, and the specifics of its organization powerfully influence our lives. From consumer choice to sexual intercourse, from tool use to chronic pain to the process of healing, the genes, cells, and neural circuits involved in the sense of touch have been crucial to creating our unique human experience.

The transcendence of touch resides in the details, and, of course, these have been sculpted over the course of millions of years of evolution. They're in the dual-function receptors in our skin that make mint feel cool or chili peppers hot. They're in the dedicated nerve fibers in our skin that predispose us to like a soft caress (but only if it moves at the proper speed across the skin). And they're in the brain's specialized centers for emotional touch, without which an orgasm would seem more like a sneeze—convulsive, but not compelling. And lest we begin to think that everything's hardwired and predetermined, these same emotional touch centers are neural crossroads where sensation and expectation collide, allowing for powerful effects of life history, culture, and context. Activity in these brain regions determines whether a given touch will feel emotionally positive or negative, depending upon the context in which it occurs. Imagine a caress from your romantic partner during a sweet, quiet connected time versus one that is administered right after he or she has said something deeply offensive. Similarly, these regions are where the neural signals engaged by the placebo effect, hypnotic suggestion, or even mere anticipation can act to dull or enhance pain. There is, in fact, no pure touch sensation, for by the time we have perceived a touch, it has been blended with other sensory input, plans for action, expectations, and a healthy dose of emotion. The good news is that these processes are no longer entirely mysterious. Recent years have seen an explosion in our scientific understanding of touch, revealing new ideas that help explain our sense of self and our experience of the world. So let's dive in. The water's not so cold once you get used to it. It will feel great.

CHAPTER ONE

THE SKIN IS A SOCIAL ORGAN

Warsaw, 1915

Solomon Asch was brimming with excitement. At the age of seven he had been allowed to stay up past his usual bedtime for his first Passover Seder. In the warm glow of the candles, he watched his grandmother pour an extra glass of wine that didn't match a place setting.

"Who's that glass for?" Solomon asked.

"It's for the prophet Elijah," explained an uncle.

"Will he really come inside and have some wine?"

"Certainly," the uncle replied. "You just watch when the time comes, and we open the door to let him in."

The extended family gathered around the table and read from the Haggadah, which tells the story of the Jews' liberation from slavery in Egypt in the time of Moses. Following the teachings of the Talmud, prayers were intoned, wine was drunk, parsley was dipped in salt water, and a festive meal was consumed while reclining in the manner of free people of the ancient world. After the meal, as tradition dictates, the front door was opened to admit the prophet. A moment later, Solomon, primed with expectation and inspired by the Passover ritual, saw the meniscus in the wineglass drop just a bit, as if Elijah had taken a single sip before slipping out the door to visit other Jewish families.

Solomon Asch emigrated with his family to New York City at the age of thirteen and soon learned English by reading the novels of Charles Dickens. As he grew older, he became fascinated with psychology, particularly social psychology, and earned his doctorate in that field in 1932 at Columbia University (figure 1.1). Years

Figure 1.1 Solomon Asch, a leader in the fields of social and Gestalt psychology. This photo was taken in the 1950s, when Asch was a professor in the department of psychology at Swarthmore College. He died in 1996, at the age of eighty-eight. Used with permission of the Friends Historical Library at Swarthmore College.

later he credited his interest in the field to his experience on that boyhood Passover night. How could the collective faith of the Seder celebrants lead him to believe in something like the prophet's sip of wine, which was demonstrably impossible? This was not just an academic question. With the rise of Hitler and Nazism in Europe, Asch became particularly concerned with two related sociopolitical questions that would hold his attention throughout his career: How can the social world shape our beliefs in the face of clear contradictory evidence? And how do we come to form rapid decisions about another's character? He wrote, "We look at a person and immediately a certain impression of his character forms in us. A glance, a few spoken words are sufficient to tell us a story about a highly complex matter. We know that such impressions form with remarkable rapidity and with great ease. Subsequent observation may enrich or upset our first view, but we can no more prevent its rapid growth than we can avoid perceiving a given visual object or hearing a melody."[1]

Asch wanted to know if there were underlying principles that guided this quick formation of character impressions. After all, everyone we encounter presents us with an array of diverse characteristics. One person is courageous, intelligent, with a ready sense of humor, and swift in his movements, but he is also serious, energetic, patient, and polite. Another is slow, deliberate, and serious, but has a fast temper when provoked. How do such perceived characteristics come together to form an overall impression of an individual and enable us to extrapolate and predict his behavior in various circumstances? Does each separate characteristic join together in a whole to form our perception, or does one particular characteristic, or a small cluster of them, dominate our overall impression? Crucially, how do these processes play out for public figures like Hitler, Churchill, or Roosevelt, with whom few people interacted directly?

In 1943, midway through World War II, Asch devised an experiment to begin to address these questions. He recruited subjects—mostly young women during these wartime years—from undergraduate psychology classes at various universities in New York City, such as Brooklyn College and Hunter College. "I shall read to you a number of characteristics that belong to a particular person," he told one assembled group. "Please listen to them carefully and try to form an impression of the kind of person described. You will later be asked to write a brief characterization of the person in just a few sentences. I will read the list slowly and will repeat it once: Intelligent . . . skillful . . . industrious . . . cold . . . determined . . . practical . . . cautious." A second group heard the same list with a single substitution: "cold" was changed to "warm." A sample response from someone forming an impression from the series including "cold" read: "A very ambitious and talented person who would not let anyone or anything stand in the way of achieving his goal. Wants his own way. He is determined not to give in, no matter what happens." A member of the warm group, in contrast, replied: "A person who believes certain things to be right, wants others to see his point, would be sincere in an argument and would like to see his point won." The subjects were also asked to elaborate on their impression by picking one of a pair of opposite descriptive terms (like generous/ungenerous, sociable/unsociable, humane/ruthless, strong/weak, reliable/unreliable, or dishonest/honest) to describe the "cold" and "warm" individuals. When the responses were analyzed and appropriate statistical tests applied, it became clear that the warm/cold distinction was very significant. The person described as warm was more often rated as generous, sociable, and humane, while the cold person was viewed as ungenerous, unsociable, and ruthless. The warm person was not, however, rated as more reliable, strong, or honest, indicating that the "warm" descriptor did not confer an across-the-board positive shift in impression. Rather, perceiving someone as warm indicates a specific constellation of traits: helpfulness, friendliness, and, most important, trustworthiness. Simply put, warm people are not identified as threats.[2]

Subsequent experiments and observations outside the laboratory by Asch and many others have shown that the warm/cold dimension is the strongest component of both the first impressions of individuals (the second strongest being competent/incompetent) as well as of group stereotypes when evaluated across many countries and cultures.[3] Why do we respond so naturally to the linguistic metaphor of the warm individual? It is likely that this metaphor has deep biological roots. We often use terms familiar from our sensory experiences to organize abstract psychological concepts. In both our individual lives and our human evolutionary history, the physical sensation of warmth has been associated with security, trust, and an absence of threat, mostly through the experience of a mother's touch.[4]

Asch's model of impression formation suggests an obvious

question: Does physical warmth relate to metaphorical warmth? Specifically, in adults, does the mere tactile experience of warmth on the skin activate feelings of interpersonal warmth that can transfer to our evaluation of an unknown person? To investigate this question, Lawrence Williams and John Bargh, of the University of Colorado and Yale, respectively, devised a clever experiment. Subjects were met in the lobby of the psychology building by an employee of the experimenters. This employee—who, crucially, was blind to the intent of the study—was carrying an awkward load: a cup of coffee, a clipboard, and two textbooks. During the elevator ride up to the lab on the fourth floor, she casually asked each subject to hold her coffee cup while she recorded the subject's information on a form attached to the clipboard. She then took the coffee cup back and delivered the subject to the experimenters. In some cases the cup contained hot coffee, and in others, iced coffee. When the subjects arrived in the lab they were immediately given a personality assessment questionnaire like that used in Asch's original 1943 study but without the warm/cold terms (e.g., person A was described as intelligent, skillful, industrious, determined, practical, and cautious). They were then asked to rate this fictional person on ten traits using the previously described method of opposites (humane/ruthless, dishonest/honest, and so on). They found that subjects who had held the hot coffee cup perceived the target person as being significantly warmer (humane, trustworthy, friendly) than those who had held the cup of iced coffee. Remarkably, the brief experience of physical warmth on the skin of the hands did indeed promote interpersonal warmth.[5]

Is the influence of incidental tactile experience on our evaluation of an unknown person unique to warmth, for which we have an unusually strong positive emotional association, or does it apply more broadly to touch sensations in general? Could other skin sense experiences subconsciously influence our impressions about unrelated people and situations? Guided by rich tactile metaphors in the English language, like "weighty matters," "the gravity of the situation," "smooth negotiation," and "hard bargainer," John Bargh, now joined by Joshua Ackerman and Christopher Nocera, set out to test this broader hypothesis.[6] First, they had passersby evaluate a single job candidate by reviewing a résumé affixed to either a light or a heavy clipboard (340 grams versus 2,041 grams, the latter being the weight of a typical medium-sized laptop computer). Subjects given the heavy clipboard rated the candidate as significantly better overall and as showing more interest in the position. The tactile experience of the weighty clipboard subconsciously caused the job applicant to be perceived as having significantly better performance and more serious intent. Importantly, the weighty clipboard did not influence subconscious impressions generally: For example, the candidate was not rated as more or less likely to get along with coworkers. Rather, the heavy clipboard specifically conferred gravitas.[7]

Bargh's group moved on to explore texture, influenced by metaphors like "I had a rough day." In this study, passersby were first asked to complete a simple jigsaw puzzle, one group being given a puzzle in which the pieces were covered in sandpaper, and the other group an identical puzzle with smooth pieces. All subjects were then asked to read a passage describing an interaction in which the social valence of the situation was deliberately ambiguous. When subjects were asked to rate the quality of the interaction in the passage, the rough-texture subjects scored it as significantly more adversarial (versus friendly), more competitive (versus cooperative), and more like an argument than a discussion. The physical experience of a rough texture changed evaluations of a social interaction to make it seem more metaphorically "rough."[8]

Finally, a similar study was conducted using hard versus soft as the tactile manipulation. Here the psychologists disguised the touch experience by embedding it in the ongoing patter of a magic act. Passersby were asked to watch a magic trick and guess its secret. As a part of the show, subjects were asked to examine an object to be used in the trick, ostensibly to confirm that it had not been tampered with. Some subjects fondled a soft swatch of blanket; others, a hard block of wood. The magic act was then postponed while the subjects were asked to read the same deliberately ambiguous social interaction passage used in the rough/smooth study, with the difference being that they were told that the interaction was between an employee and a boss. When asked to evaluate the employee, subjects primed with the wood block were significantly more likely to rate him/her as rigid and strict, consistent with hardness as a metaphor for an unyielding, unemotional personality. (And, no, sadly, the subjects never actually got to see the magic trick.)

The fact that even incidental touch experience can influence our impressions of people and our social interactions can be more than a little disconcerting. How would that beautiful, intelligent woman I was chatting up at the Caffe Med in Berkeley in 1983 have reacted if she had had her hand wrapped around a hot cup of coffee instead of that cold Italian soda? And what about that weird department chairman who compulsively squeezed a hollow rubber ball while he interviewed me for an academic job? If he had been toying with a letter opener, would he have tended to regard me as sharper, or merely hard and unyielding?

~~~~~~

While the Bargh incidental touch studies were well designed and their findings are useful and interesting, they do have serious limitations. Most important, they do not capture impressions as they occur in real-world situations. Survey-based experiments require that subconsciously formed impressions be made conscious and explicit, and the responses of the subjects must conform to the experimenter's predetermined scales of measurement. This is an unnatural situation. In our daily lives we are continually forming ideas about people and situations but are not simultaneously making mental checklists assessing them on the basis of "humane versus ruthless" or "discussion versus argument." That's why it's important to investigate the social roles of touch in everyday contexts.

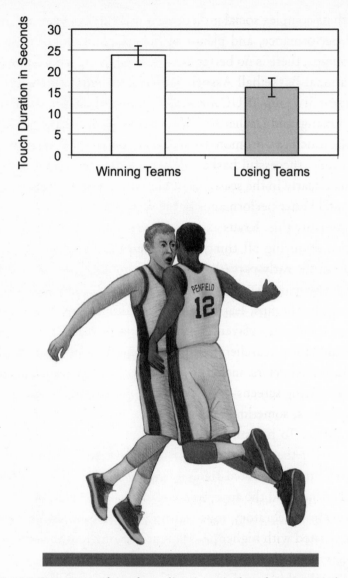

Figure 1.2 Interpersonal touch predicts increased performance in the NBA. Top: Bar graph shows total celebratory touch duration scored in a single early season game for winning and losing teams in the five games that followed during the 2008–9 NBA season. From M. W. Kraus, C. Huang, and D. Keltner, "Tactile communication, cooperation and performance: an ethological study of the NBA," *Emotion* 10 (2010): 745–49. Published by the American Psychological Association; reprinted with permission. Bottom: Basketball players engaging in an airborne chest bump.

With its complex social milieu, clear measures of team and individual performance, and plenty of butt slapping, high fives, and chest bumps, there is no better tactile/social living laboratory than the National Basketball Association (NBA). This was the insight of a research group at UC Berkeley composed of Michael Kraus, Cassy Huang, and Dacher Keltner. They reasoned that since interpersonal touch can promote trust and cooperation, two important factors for a successful basketball team, increased touch between teammates early in the season should predict more cooperative behavior and better performance as the season unfolds.

To measure this, Kraus and coworkers first watched videotapes of games featuring all thirty NBA teams (294 players were involved[9]) in the early part (the first two months) of the 2008–9 season. They scored the occurrence, type, and duration of celebratory touching (fist bumps, leaping shoulder bumps, high tens, and so on) that followed a player's making a successful shot. For these same games the researchers also produced ratings using an index of cooperative behaviors, including talking to teammates, passing the ball, and setting screens—that is, behaviors that showed a reliance on teammates, sometimes at the expense of one's own individual performance. To measure individual and team performance over the course of the season, they turned to statistics maintained by the NBA and distributed freely on its Web site.[10] When the data were crunched and the appropriate statistical tests run, there was a clear result: Celebratory touch during games early in the season was associated with higher performance throughout the season for both individual players and teams (figure 1.2).

But might this association come about trivially? For example, what if the best players and teams simply score more often and therefore have more cause to celebrate with touch? That would change the interpretation of the touch-performance correlation. To address this possibility, Kraus and his colleagues applied a statistical correction to adjust for the total numbers of points scored, but the touch-performance correlation still held strongly for both teams and individuals. But what if teams that were predicted at the outset of the season to do well (in polls of coaches or sportscasters) were more optimistic, and it was this factor that led to more celebratory touching and better performance? Again, the predictive power of early-season touch on performance remained after applying a statistical correction based on early-season predictions, as well as another correction based on player status (using salary as a proxy measure).

Finally, when the cooperation scores were analyzed, it emerged that cooperation largely accounts for the relationship between touch and enhanced performance. While investigations of this type cannot prove causality, the correlations in this study strongly indicate that, at least within the context of professional basketball, brief celebratory touch enhances individual and group performance and does so by building cooperation.

<center>~~~~~</center>

But for those of us who don't play in the NBA, what social functions are served by interpersonal touch? Is social touch always intended to foster trust and cooperation? A good way to start approaching these questions is to examine certain of our closest primate cousins— baboons, chimpanzees, bonobos, and vervet monkeys. These species live in large social groups and have many eyes and ears directed at the perimeter of their territory to detect approaching trouble and to keep themselves safe from predators. There is also strength in numbers, for while an adult leopard can almost always prevail in a fight with a single baboon, groups of baboons have been reported to drive leopards up a tree, and even, occasionally, to kill them. Many of these large social units are situated in places with ready access to food. The stability afforded by a low risk from predation and plenty of food gives baboons lots of free time to engage in complex social lives. For instance, Robin Dunbar reports that gelada baboons living in the Ethiopian highlands spend up to 20 percent of their waking hours fussing over the skin and fur of other geladas. That's a huge investment of time in grooming behavior. While grooming is useful to remove dead skin, parasites, tangled hairs, and bits of plant material, the time that gelada baboons (and many other primate species) devote to this task is far out of proportion to the health benefits that accrue from having well-tended skin and fur. The main reason for extended grooming is social, not dermatological (figure 1.3).

Figure 1.3 An adult male gelada baboon (*Theropithecus gelada*) is being groomed by a younger male. This behavior is key to building lasting social bonds and forming alliances.

Gelada live in large troops, typically numbering one hundred to four hundred individuals, but within each troop are many smaller social units: harems, consisting of four or five females, their young offspring, and a single breeding male. When young gelada reach puberty, the males in a harem will depart to join a bachelor group,

but the females will remain, assuring that the social core of the harem is a group of related females: mothers, sisters, aunts, and female cousins. These harem females form a loyal and long-lasting
coalition that is strengthened and maintained by extensive grooming.[11] Their sisterly solidarity is displayed in a number of ways, but
none is more amusing than when a member of the coalition is
threatened by a domineering male. The single breeding male in the
harem must constantly police his females to keep them from having sex with the younger males of the bachelor groups who are always loitering nearby. In addition to scaring off the bachelors, the
breeding male will often attempt to intimidate a straying female of
his harem with a charge and threat display (panting and teeth
grinding). At this point, her relatives race to the rescue, ganging up
to chase off the breeding male. But within the female coalition, not
all relationships are equal, and some bonds are stronger than others: In an interharem squabble, a female will ally herself with her
most avid grooming partner.[12]

In these primate social groups, grooming is as laden with social
meaning as choosing their lunch table companions is for high
school students. Mothers groom their offspring; mating partners
groom each other; friends groom friends, in both male and female
pairs. And, just as in high school, the higher-status primates receive more grooming attention than they give. A network of loyalty is created and reinforced by grooming, so coalition members
are more likely to come to the aid of one of their group when he or
she is confronted by another in the harem or in the larger troop, or
even by a predator. Using recording and playback equipment in the
field, it has been shown that chimpanzees and macaque monkeys
are more likely to respond to a distress call (thereby putting themselves in danger) when the call was recorded from an animal with
which they recently groomed.

A young male chimpanzee or baboon may seek to curry favor
with the breeding male by grooming him or by establishing a
grooming-based alliance with another young male to seek to

overthrow the breeding male. If he succeeds in overthrowing the king, it is to his advantage to make a gesture to the deposed monarch to reduce the chance that he will attack in a bid to regain his former status. If he is clever, he might even get the former king to become an ally in fending off other males during the days of transition. If the deposed king believes that he cannot regain his status, it is to his advantage to reconcile as well, as he would like to remain in the group to protect his last batch of offspring, even if his breeding days are now over. There's a ritual for this reconciliation: The victor presents his rear end to the deposed male, who then reaches through the new monarch's legs to lightly touch his penis. With those formalities concluded, the pair grooms each other like long-lost friends to seal the deal.

~~~~~~~

So the situation is not dissimilar for human NBA players and certain nonhuman primates like gelada baboons: Social touch tends to reinforce cooperation and loyalty. Human and nonhuman primates alike use grooming and other forms of social touch to soothe, reconcile, form alliances, reward cooperative actions, and reinforce bonds of kinship and friendship. Has this type of behavior appeared only in the primate lineage, or are there traces of it in other animals as well?

There is at least one notable example of social grooming and cooperation in a nonprimate. The common vampire bat, *Desmodus rotundus*, takes wing at night to feed on the blood of living mammals, most commonly horses, burros, cattle, and tapirs. This is their only food source, because their narrow throats cannot accommodate solid food. If the animal they are feeding upon has fur, they will use their canine and cheek teeth to carefully shave away a patch prior to piercing the skin with sharp upper incisors to start bleeding. The bats' saliva contains an anticoagulant compound that keeps the blood flowing for the twenty to thirty minutes of lapping needed to consume a meal. (Sometimes another bat will wait patiently to feed at the same wound.) An adult female vampire bat typically weighs about 40 grams but can consume a blood meal of 20 grams before flapping away, laden and sated. However, vampire bats have a very high metabolism, and if they fail to find a meal on two successive nights, they will lose about 25 percent of their body weight and be near death.

In one part of their range, in northwestern Costa Rica, vampire bats live in hollow trees in groups of eight to twelve. Gerald Wilkinson and his colleagues from the University of Colorado observed these bats in their tree roosts every day for many months.[13] They found that the animals were more likely to groom each other if they were closely related or if they were frequent roost mates. Grooming also promoted cooperation of a particular kind: A bat that was just groomed was more likely to share a blood meal with its groomer through regurgitation (figure 1.4). In fact, the grooming seemed to function as a type of solicitation for food sharing. By begging food

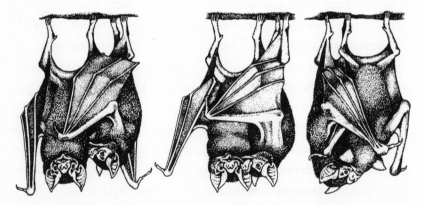

Figure 1.4 A hungry bat solicits regurgitated blood by grooming. The grooming starts with the hungry bat licking the potential donor under her wing (left), and then licking the donor's lips (center). If receptive, the donor responds by regurgitating blood (right). Only bats that are close relatives or exhibit long-term roosting associations provision blood to one another. Illustration by Patricia J. Wynne; used with permission. This drawing first appeared in G. S. Wilkinson, "Food sharing in vampire bats," *Scientific American* 262 (1990): 76–82.

from a roost mate who has just returned from feeding, a bat can fend off starvation for one more night, and so have a chance to find its own blood meal. In the vampire bat world, this is what counts as a win-win deal: I'll groom you, and you'll vomit blood down my throat. Next time, in the spirit of reciprocity, maybe I'll do the same for you.

~~~~~~

We've seen a lot of evidence that social touch can promote trust and cooperation. Underlying our interpretation of these findings has been the presumption that all these mammals—geladas and humans and bats alike—shared an early experience of a mother's touch that caused them to associate warm, gentle social touch with safety. What happens when this early maternal experience is lacking?

In the late 1950s Seymour Levine and his coworkers at the Ohio State University Health Center studied the role of early postnatal life on the development of personality, particularly stress responses. They bred Norway rats in the lab, and soon after birth would pick three pups from a litter (which typically consists of ten to twelve pups) and handled them gently for fifteen minutes. This procedure would be repeated every day with the same three pups until they were twenty-one days old. When these handled pups grew into adults, they exhibited a set of positive behavioral traits: They were less fearful, more likely to explore novel environments, and less responsive to stress when compared with their nonhandled littermates. When blood samples were taken from adult rats that had been handled as pups, it was found that brief exposure to stress in adulthood evoked in them a lower secretion of the stress hormones ACTH and corticosterone.[14]

These initial studies did not address the means by which handling actually triggered the behavioral and hormonal changes in stress response. Levine suggested that it was not the handling per se that caused the changes but rather the subsequent behavior of the mother rat. When the pups were returned to their home cage after handling, they emitted ultrasonic cries, and in response the mother rat doubled her rate of licking and grooming them. This increased tactile attention persisted throughout the period in which the pups were handled.

While the behavior of mother rats is fascinating for its own sake, we'd ultimately like to know if the lifelong reduction in stress responses of human-handled rat pups is relevant for human development. Interest in this question was sparked by a key set of experiments from a research team at McGill University headed by Michael Meaney. These revealed that if you examine many rat mothers (all Norway rats of the same laboratory strain, called Long-Evans), some lick and groom their pups a lot, while others do less. In fact, the most attentive mothers spent about threefold more time licking and grooming than the least attentive mothers. Furthermore, human handling of the pups could normalize this variation: Following handling, the low licking-grooming mothers increased their pup licking-grooming time to match that of their most attentive peers.[15]

When the pups of low licking-grooming mothers grew up, they had impaired spatial learning and more fearful behavior compared to those of high licking-grooming mothers. They were also less likely to explore a new environment or try a new type of food.[16] To be a bit anthropomorphic about it, they were wimps. Their fearful behavior can be related to stress-hormone signaling: Adult rats that were offspring of low licking-grooming mothers had lifelong increases in hormonal responses triggered by stress (figure 1.5).

What conclusion should we take from the correlation between low licking-grooming mothers and increased stress responses in their pups? Does low licking-grooming behavior cause these effects, or is it merely correlated with them? Might low licking-grooming mothers pass these traits on to their pups genetically? And there was another twist in these experiments' findings: In an echo of humans, where poor parenting is often observed across generations, when female pups of low licking-grooming mothers

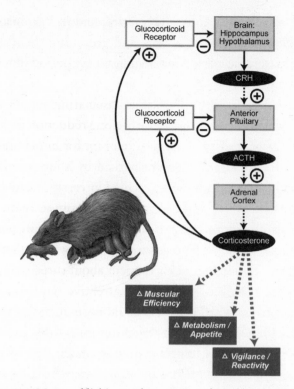

Figure 1.5 Maternal licking and grooming of newborn rat pups produces life-long changes in stress hormone signaling. Stress produces a cascade of hormonal responses that begin in a region at the base of the brain called the hypothalamus, which secretes a hormone called corticotropin-releasing hormone (CRH). CRH activates the anterior portion of the pituitary gland, which in turn secretes another hormone, adrenocorticotropic hormone (ACTH), which passes throughout the bloodstream to stimulate the adrenal gland. Then the adrenal gland releases the hormone corticosterone, which has many effects on the body, including regulation of muscular efficiency, metabolism, electrolyte balance, appetite, and vigilance. Corticosterone also binds glucocorticoid receptors in the brain to form a negative feedback loop, suppressing production of CRH. This entire stress-signaling pathway is called the hypothalamic-pituitary adrenal (HPA) axis. Pups of low licking-grooming mothers grow up to have increased levels of ACTH and corticosterone following brief mild stress. (Adult rats were confined to a plastic tube for twenty minutes, following which blood samples were taken.) The brains of these pups also have fewer glucocorticoid receptors available to bind circulating corticosterone, thereby blunting the negative feedback loop and further increasing stress hormone effects.

grew up, they were much more likely to become low licking-grooming mothers themselves.

In human behavioral science, disentangling nature-versus-nurture problems often involves studies in which twins are adopted into different families. It's possible to do something similar with rats. Two pups from the litter of a low licking-grooming mother were removed from her within twelve hours of birth, marked with a Sharpie for identification, and then cross-fostered by adding them to the litter of a high licking-grooming mother. When the pups grew up, they had reduced behavioral and hormonal stress responses as compared to their nonadopted siblings. The cross-fostered female pups were also much more likely to grow up to become high licking-grooming mothers themselves. Conversely, pups adopted in the other direction, from high to low licking-grooming mothers, had increased stress responses, and the female pups tended to grow up to become low licking-grooming mothers.[17] These results, together with the beneficial effects of human handling on the pups of low licking-grooming mothers, argue for behavioral rather than genetic transmission of stress reactivity. But the effects of high licking-grooming behavior must somehow be changing the brains and hormonal systems of the pups, so the effects are still fundamentally biological. In fact, we now know some of the biochemical details of how maternal licking-grooming persistently modifies gene expression to underlie behavioral transmission across generations. These "epigenetic signals" are some of the molecular events where nature and nurture meet.[18]

If raising pups that are resilient to stress is good, then why don't all the rat mothers engage in lots of licking and grooming, thereby giving their pups an advantage as they make their way in the world? This type of selection can occur even if the mode of transmission is behavioral, not genetic: If the pups of low licking-grooming mothers are at a disadvantage in survival and reproduction, then wouldn't high licking-grooming behavior come to dominate? The answer is complex and not entirely clear. Since wild Norway rats inhabit a great variety of ecological niches, from city dumps to meadows to forests, they must contend with a broad range of ecological demands, including different predators, food sources, and weather. Michael Meaney and his coworkers have suggested that in some ecological niches, like those with scarce food or abundant predators, it might be advantageous to have a high degree of stress reactivity, as imparted by a less-attentive mother rat: When confronted with the constant danger of being eaten or starving, it can be adaptive to be high-strung. One simple idea about how this might come about is the all-too-familiar work/home-life balance: In an echo of human societies, mother rats that have to range far and wide for food leave the nest more often and thereby have less time to lick and groom their pups.

~~~~~~

What does the connection between maternal tactile stimulation and stress responses in rats tell us about other species? Let's have a look, first down, and then up, the phylogenetic tree. The tiny soil-dwelling, bacteria-eating roundworm called *C. elegans* is about 1 millimeter long when it reaches adulthood, about three days after hatching. This creature is a favorite of biologists because it is easy to grow in the lab, it has a fast generation time, and it's transparent. We now have a complete map of its adult nervous system, which consists of 302 neurons (compared with about 500 billion in the human brain). Only six of them are touch receptors embedded in the body wall. These touch-sensor neurons provide information that triggers the worm to swim forward or backward, depending upon what it encounters (soil particles, liquid surface tension, another worm). When newly hatched worms were allowed to develop together in groups of thirty to forty in a laboratory dish filled with nutrients, they attained their full length, similar to wild worms collected from soil samples. When these adult, colony-raised worms had their touch receptors stimulated by a tap on the edge of the dish wall, they typically reversed direction and began swimming backward. However, when a worm egg is isolated in a dish, hatches, and grows up in isolation from other worms, it will not attain its potential mature length and will respond more weakly to taps to the dish wall. The isolated worms tend to just keep swimming straight ahead, as if they do not feel the vibration.

Catharine Rankin and her colleagues at the University of British Columbia found that they could completely reverse the developmental deficits in both body length and touch sensation with a surprisingly crude maneuver: placing a dish containing a worm within twenty-four hours of its hatching in a padded box and repeatedly dropping the box from a height of two inches onto a table thirty times over the course of a few minutes.[19] This treatment also reversed some biochemical and structural changes in the six sensor neurons thought to impair their ability to communicate touch events to other neurons. Even in an organism as simple as the worm, with no mother involved in child rearing and only six touch-sensitive neurons, tactile stimulation is important for development of the body and the nervous system, with effects persisting into adulthood.

In humans, touch is thought to be the first sense to become functional in utero, at about eight weeks' gestation. At that point, the human fetus is about 1.5 centimeters long, weighs about a gram, and is exhibiting its first brain activity. Touch reception continues to develop from reflexive to intentional behavior as pregnancy progresses. I have fond memories of watching my unborn twins on the ultrasound monitor responding to each other's kicks and blows during the last trimester of pregnancy. Jacob would stomp on Natalie's head, and she would respond with a staccato jab to his belly. It looked for all the world like fetal martial arts.

Once human children are born, most mothers and fathers give them sufficient tactile attention. Children do not suffer lasting health problems because they didn't receive daily baby massages, listen to Baby Einstein CDs, or get entertained by motorized crib mobiles. Studies that have investigated the role of touch sense in child development have, however, examined cases of drastic touch deprivation, as occurred in understaffed orphanages or with premature babies isolated in incubators. There have now been many such studies, and the results are clear: Severely touch-deprived infants and preemies have a broad range of developmental problems, ranging from impaired growth, increased vomiting, and compromised immune system function to slowed cognitive and motor development and the emergence of attachment disorders. As in rat pups, these effects were not limited to early life. Persistent touch deprivation of infants results in a significantly higher incidence of obesity, type 2 diabetes, heart disease, and gastrointestinal disease in adulthood. Adult neuropsychiatric problems are also found at a much higher rate, including anxiety, mood disorders, psychosis, and poor impulse control.[20]

Of course, one must be appropriately critical of these epidemiological studies: For example, babies raised in understaffed orphanages are likely to be malnourished and have substandard medical care as well, and are more likely to grow up poor. Likewise, premature infants have a host of developmental problems that are unrelated to touch deprivation. What's important to realize is that although studies of correlation can never be definitive as to cause, careful methods of analysis can increase confidence in their causal nature. For example, appropriate design and statistics have shown that there are strong effects of touch deprivation even when populations are normalized for nutrition, medical care, or poverty.[21]

The good news is that it doesn't take much effort to reverse the deleterious effects of touch deprivation on infants. In understaffed orphanages, twenty to sixty minutes per day of gentle massage and limb manipulation was able to mostly reverse the negative consequences of touch deprivation. The babies receiving touch therapy gained weight more quickly; had fewer infections, better sleep, and decreased crying; and progressed more rapidly in developing motor coordination, attention, and cognitive skills.

For preterm babies, one of the effective methods of delivering gentle tactile stimulation is called kangaroo care. This technique

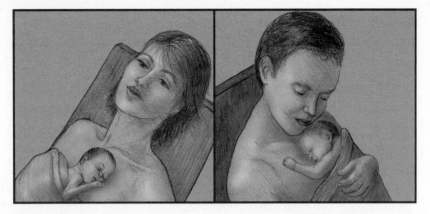

Figure 1.6 Kangaroo care for preterm infants. The baby wears only a diaper (and sometimes a hat) to maximize skin-to-skin contact. Mothers provide most kangaroo care, but fathers can pitch in as well. Today, more than 80 percent of neonatal intensive care units in the United States utilize kangaroo care, largely due to the effective proselytizing efforts of Susan Ludington, a professor of pediatric nursing at Case Western Reserve University.

was developed out of necessity in an overcrowded neonatal intensive care unit at the Instituto Materno Infantil in Bogotá, Colombia. In 1978 Dr. Edgar Rey Sanabria was dealing with a terrible 70 percent mortality rate in this unit, mostly from respiratory problems and infections. There were insufficient doctors and nursing staff and not enough incubators. Dr. Sanabria encouraged mothers to devote many hours a day to skin-to-skin contact with their preterm infants to keep them warm and to provide breast-feeding on demand. The typical position used was chest-to-chest with an upright posture, inspired by the skin-to-skin contact of a mother kangaroo with a joey in her pouch. Tactile stimulation was not the main rationale for kangaroo care but turned out to be one of its primary benefits. Introduction of this method rapidly reduced the mortality rate at Sanabria's unit to 10 percent. Kangaroo care, which is inexpensive and enormously effective, has spread and positively transformed neonatal care around the world (figure 1.6).[22] One recent study followed two groups of preemies, one that received kangaroo care for fourteen consecutive days after birth and another that received standard incubator care. Impressively, clear benefits of early skin-to-skin contact could still be seen at age ten. The kangaroo care preemies grew up to be less stress-reactive and to have improved sleep patterns, cognitive control, and mother-child reciprocity.[23]

~~~~~~~

We know from our own everyday social experiences that touch can be used together with other sensory signals to communicate a broad range of emotional intentions including support, compliance, appreciation, dominance, attention getting, sexual interest, play, and inclusion. These intentions have been documented in self-report studies in which participants were instructed to write down a brief note immediately after they touched someone or were touched themselves.[24] While self-reports have the advantage of analyzing behavior that occurs out in the world, where life is lived, they also have the disadvantage that all the wonderful multisensory and situational complexity of the real world makes it hard to tease out the precise role of touch in any given interaction.

Let's dig a little deeper. Can touch communicate specific emotions, or does it merely intensify emotions primarily conveyed by other senses, like sound or sight? Or, perhaps, the answer lies somewhere in the middle: Touch might convey emotions, but only generally, and is limited to signaling and overall tone: warmth/intimacy/trust versus pain/discomfort/aggression. Matthew Hertenstein of DePauw University and his colleagues have begun to conduct some interesting experiments to address the role of social touch in emotional communication.[25] In one study, pairs of students at a university in California sat at a table and were separated by a black curtain. They were not allowed to see each other or talk at any point. One subject, designated as the encoder, was shown a sheet of paper containing a word defining an emotion, selected randomly from a list of twelve. He or she was asked to think for a

moment about how to communicate that emotion and then to attempt to convey it by touching the bare forearm of the other subject
in any appropriate way for about five seconds. The receiving subjects,
or decoders, could not actually see the touch because their arms
were placed on the encoder's side of the curtain. After each touch,
the decoder was asked to score the intention of the encoder on a
response sheet consisting of the twelve possible emotion words
(anger, disgust, fear, happiness, sadness, surprise, sympathy, embarrassment, love, envy, pride, and gratitude) listed in random order
together, as well as the choice "none of these terms is correct." All
the touches were recorded on video and evaluated later by others
who had no knowledge of either the intended or received emotion.

When data from 106 pairs of subjects were analyzed, it was found
that the self-focused emotions of embarrassment, envy, and pride
were not effectively communicated, but that the prosocial emotions of love (encoded mostly by stroking and finger interlocking),
gratitude (handshaking), and sympathy (patting and stroking)
were decoded at levels well above chance. When examining a set of
emotions shown by other investigators to be readily conveyed by
facial expressions,[26] anger (encoded by hitting or squeezing), fear
(trembling, squeezing), and disgust (pushing away) were successfully conveyed, but happiness, sadness, and surprise were not.
Later, at a separate location, another group of subjects was asked to
score videotapes of these touches using the same list of twelve emotions plus the option "none of the above." They, too, were able to
decode love, gratitude, sympathy, anger, fear, and disgust at high
rates, but not the other emotions. The researchers concluded that
humans could indeed communicate distinct emotions through
touch and thereby argued that touch is not limited to intensifying
or shading the meaning of emotions primarily conveyed by other
senses.

As always, though, the devil is in the details. Of course, ideas
and expectations about touch differ across cultures, in specific interactions between the genders, and even in particular situations,

and these variables can influence touch communication. When the anonymous arm-touch experiment was repeated in Spain, the results were almost identical. However, when Hertenstein and co-workers reexamined their data set from the Californian anonymous touch subjects several years later, some interesting gender effects emerged.[27] When a woman attempted to convey anger to a man, he never decoded the touch intention accurately. And when a man tried to convey sympathy to a woman through anonymous touch, she was never able to interpret the message.

These anonymous-touch laboratory experiments are useful in defining the boundaries of what touch can communicate in isolation, but, of course, no one actually uses touch to communicate in comparable conditions in real life. First, most touch communication does not take place between strangers—in most cases it's a more intimate form of interaction. Second, touch in the real world is never devoid of context. We know from our own experience that the very same touch sensation can convey a very different emotional meaning, depending on the gender, power dynamic, personal history, and cultural context of the touch initiator and receiver. An arm around the shoulder can convey a variety of intentions, ranging from group inclusion or sympathy to sexual interest or social dominance.[28] And, of course, the cultural influence on social touch, particularly public touch, is enormous. In the 1960s, the psychologist Sidney Jourard observed pairs of people engaged in conversation in coffee shops around the world.[29] He methodically watched the same number of pairs in each location for the same amount of time. Jourard found that couples in San Juan, Puerto Rico, touched an average of 180 times per hour, compared with 110 times per hour in Paris, 2 times per hour in Gainesville, Florida, and 0 times per hour in London.[30] Similar differences were seen when touches were scored among people of twenty-six different nations in the departure lounges of an international airport on the West Coast of the United States.[31] Public farewell touching was most common among people born in the United States, in the Latin/Caribbean

region, and in Europe, and was much less common among people
born in northeast Asia.

<center>~~~~~</center>

The powerful effects of culture, gender, and social situation on interpersonal touch perception raise a crucial question: How can the
very same sensation (say, a brief arm-around-the-shoulder squeeze)
delivered with the identical pressure and dynamics and leading to
precisely the same signals conveyed to the brain from the skin and
muscles give rise to such different perceptions in the people experiencing it? Importantly, it's not that such sensations feel the same to
all individuals at the moment they experience them, but are then in-
terpreted differently upon reflection. Rather, they actually feel different from the very first instant they can be consciously detected.
The tactile perception of an around-the-shoulder squeeze from a
domineering boss feels fundamentally different from the same gesture received from a good friend, which in turn feels different from
that of a lover. The raw stuff of tactile sensation must be combined
with the imprints of our life experience—starting in the womb and
continuing to the present moment, soaking up culture and gender
roles and personal history along the way—to ultimately produce our
highly nuanced perception of social touch. This combination of the
past with the present has to take place within about a tenth of a second. Now our task will be to explore how the biology of skin,
nerves, and brain underlies this crucial integrative aspect of our lives
as social animals.

CHAPTER TWO

PICK IT UP AND PUT IT
IN YOUR POCKET

Even the giants of philosophy can have a bad week. Aristotle, for example, struggled mightily with the problem of human cognitive superiority. How can humans be so much cleverer than other creatures when the hawk has keener eyesight, the dog has a finer sense of smell, and cats have more acute hearing?[1] Pondering this dilemma, Aristotle came to believe that human touch sense was unusually acute, and that it was this acuity that accounted for the superior intellect of our species:

> The human being is left behind by many of the animals, but with respect to touch he is precise in a way that greatly surpasses the rest, and this is why he is the most intelligent of the animals. A sign of this is that within the human race, being naturally well or badly endowed with intelligence depends on the organ of this sense and not on the others, for those with tough skin are badly equipped by nature for thinking, but those with tender skin are well equipped.[2]

Research on the biology of touch has not supported either the basis or the inference of Aristotle's argument. In fact, our tactile perception is not the most sensitive of all animals. Furthermore, among humans there appears to be no correlation between intelligence and either skin softness or acuity of fine touch perception. How did the philosopher get this so wrong? It's likely that Aristotle had an agenda based upon social class: In his view, slaves and others whose hands were roughened by manual labor were clearly not as bright as philosophers and nobles, the soft-skinned elite.

What Aristotle didn't know is that we (and other animals) have an array of many types of touch sensors in the skin, each a beautiful, specialized micromachine sculpted by evolution to extract different aspects of information about our tactile world. The nerve fibers that relay the information from these touch sensors to the spinal cord are mostly dedicated to a single class of sensation, one for texture, another for vibration, another for stretching, and so on. When we use tactile information to play the violin or make love or take a sip of coffee, we don't have to think about the array of various sensors in the skin. The streams of information from these sensors are blended and processed in our brains, so that by the time we have conscious access to them, they are in the form of a unified and useful percept. Moreover, touch information is subconsciously combined with inputs from vision, hearing, and proprioception (a sense of where our bodies are located in space that comes from nerve endings in our muscles and joints) to give rise to rich, nuanced perception.

~~~~~~

The skin is an interface between our internal and external worlds and is, by virtue of this topology, the locus of touch. In addition to allowing touch information in, it has to keep a lot of dangerous things out. Skin serves as a barrier to repel noxious stressors like parasites, microbes, mechanical and chemical insults, ultraviolet radiation, etc. To aid in this task, it has its own specialized branch of the immune system and secretes its own hormones.[3] The skin of an individual human is surprisingly big. If, in the tradition of the splatter film, I were to fall victim to a leering psychopathic murderer and she were to carefully flay my corpse, my resultant hide would weigh about as much as a bowling ball (14 pounds), making it the largest organ in the human body. To get a stomach-lurching sense of the skin's total surface area, imagine taking delivery of nine family-sized pizza boxes and laying them in a three-by-three grid on the floor.[4]

There are two basic types of skin: hairy and hairless. The medical term for skin without hair is glabrous,[5] and you might think that glabrous skin can be found in lots of smooth places: Keira Knightley's cheek, for example. But if you were to look carefully, you'd find that this lovely soft patch of her face is actually covered with many fine, short, light-colored hairs called vellus hairs. Other apparently smooth places on the human body, like the inner bicep or inner thigh, also have them. These soft vellus hairs have an essential wicking function, drawing perspiration away from the skin and thereby increasing the efficiency of evaporative cooling. The only true glabrous skin is found on the palms of the hands (including the inner sides of the fingers), the soles of the feet, the lips, the nipples, and portions of the genitals.[6] In women, the skin of the labia minora and the clitoris is glabrous, but the skin of the labia majora is hairy. For men, the foreskin and the skin covering the head of the penis, called the glans, is glabrous, but the shaft of the penis is hairy (even on glistening, chemically depilated porn actors).

Hairy and glabrous skin have the same general structure. Imagine a two-layer cake with the top layer divided into sublayers (figure 2.1). Both types of skin have an outermost sublayer of flattened dead skin cells, called the stratum corneum, and three underlying sublayers, each of which contains a mixture of several types of living cells, including keratinocytes, Langerhans cells (which are part of the immune system), and melanocytes. The melanocytes manufacture granules of the pigment melanin, which is the main determinant of skin color. Together, these four sublayers comprise the epidermis. The cells of the epidermis are continually regenerated, with new ones created by cell division in the deepest sublayer, gradually migrating upward. In this migration, the cells flatten as they are pushed up from below, and their internal structure breaks down, leaving, in the stratum corneum, tough, pancake-shaped cell husks that are ultimately shed from the skin surface. In this way, the epidermis is completely renewed about every fifty days. Below the epidermis is another layer, the dermis, which contains nerves, blood vessels, sweat glands, and a dense network of elastic fibers.

The key structural differences between glabrous and hairy skin are shown in figure 2.1. Hairy skin has both fine, pale vellus hairs and also the longer and thicker, more visible guard hairs. The epidermis of glabrous skin tends to be thicker than that of hairy skin.[7] It also has a different shape—wavy, rather than flat. On the surface of the skin, we know these wavy ridges, called papillary ridges, as

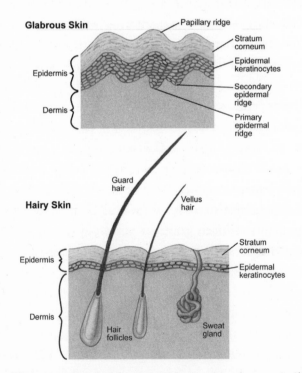

Figure 2.1 The structure of glabrous and hairy skin share a similar layered pattern but have some important differences as well.

fingerprints (and toe, palm, and sole prints, as well). The internal face of the wavy epidermis forms partially complementary structures called primary and secondary epidermal ridges, and we can think of them as inward-facing fingerprints.

As a symbol, fingerprints have a deep emotional and spiritual resonance. There's something fascinating about them—they're an external mark of human individuality written in an obscure artistic code. Fetal fingerprints begin to form at about age twenty-six weeks and have already achieved their adult shape at birth. In the tradition of the Dineh people (also known as the Navajo), the Spirit Winds, a kind of life force, are said to emanate from fingerprints:

> There are whorls here at the tips of our fingers. It is the same way on the toes of our feet and Winds exist on us here where soft spots are, where there are spirals . . . These Winds sticking out of the whorls at the tips of our toes hold us to the Earth. Those at our fingertips hold us to the Sky. Because of these, we do not fall when we move about.[8]

That wonderful description is both moving and evocative. But from a biological point of view, what function do fingerprints (and palm, toe, and sole prints) serve? A long-standing hypothesis holds that they aid in climbing and grasping, but this notion has been challenged. When the friction between a fingertip and a smooth, dry surface was measured, it was found, counter to expectations, that fingerprints reduced grasping efficiency by about 30 percent. However, when a surface is wet or rough, fingerprints increase friction and stabilize grasp. In that respect they resemble automobile tires: Race cars, which encounter only smooth, dry racetracks, are fitted with smooth tires to maximize the contact area between tire and roadway and thereby provide maximum grip. Passenger cars, in contrast, are typically driven over wet and uneven surfaces, and in these conditions, tires with grooves cut to channel water away from the contact area are superior.

Fingerprints are not uniquely human: Gorillas and chimpanzees have them as well. Nor are they unique to primates. Fingerprints appear in a scattershot fashion throughout mammalian species. In Australia they are present in koalas, but not in a close relative of koalas called the hairy-nose wombat, or in another arboreal creature, the tree kangaroo.[9] Fingerprints are found in fishers, a type of North American weasel, but not in some other members of the weasel family. At present it's not certain if the presence of fingerprints in a particular species is related to its grasping behavior. For all their symbolic significance, we still don't really know what fingerprints are for.

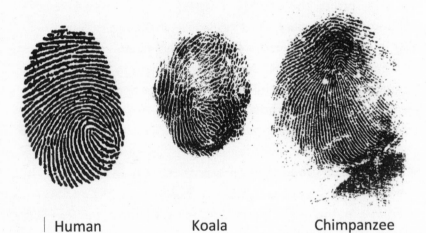

| Human   Koala   Chimpanzee

Figure 2.2 Human, koala, and chimpanzee fingerprints are almost indistinguishable. The evolutionary paths of humans and koalas diverged at least 70 million years ago, yet humans and koalas share fingerprints, while other species that are close relatives of koalas do not. Reprinted from M. J. Henneberg, K. M. Lambert, and C. M. Leigh, "Fingerprint homoplasy: koalas and humans," *NaturalScience* 1, article 4 (1997), with permission of Heron Publishing. Thanks to Dr. Maciej Henneberg, University of Adelaide, Australia.

I can still hear my mother's voice: "Get out of the bath. You're turning into a human prune!" Many people believe that the wrinkly finger and toe pads one gets from prolonged exposure to water are the result of a passive process in which water is gradually absorbed by the dead skin cells of the stratum corneum. However, this was shown to be false as early as 1936.[10] The key observation regarding this phenomenon is that finger and toe wrinkling do not occur if the electrical signals flowing from the spinal cord to the skin are interrupted by cutting the nerve or applying a drug that blocks the nerve's signals, manipulations that have no effect on the stratum corneum. In particular, the wrinkling response requires a branch of the subconscious autonomic nervous system called sympathetic outflow.[11]

So what, if anything, is the purpose of the wrinkling response? Mark Changizi and his colleagues at 2AI Labs have suggested that, like fingerprints, the wrinkles function as rain treads to increase traction on wet surfaces. They note that the wrinkling response is also found in macaque monkeys and chimps and suggest that it may be an adaptation of primates to wet, slippery conditions.[12] In support of this hypothesis, Kyriacos Kareklas and his colleagues at Newcastle University showed that subjects with wrinkled fingers were able to transfer wet marbles from one container to another at a significantly higher speed than subjects with unwrinkled fingers. Wrinkled fingers conferred no advantage in handling dry marbles, however.[13]

~~~~~

How are the specialized touch sensors arranged in the skin, and how does this arrangement influence our tactile experience? This turns out to be a big question. To explore it, let's take an everyday manual task and break it down into tiny steps. You're running late for a movie and are relieved to find a parking place in the crowded streets around the theater. Standing in front of the old-style mechanical parking meter, you see that it takes only quarters. Fishing in your pocket of loose change and miscellaneous junk, you feel around until you find a quarter, extract it, and insert it in the meter's slot. You then grip the handle and twist. As you twist you feel that satisfying action of the ratchet mechanism in the meter engaging, the *ker-plink* vibration of the quarter falling inside, and, finally, the twisting force as the handle counterrotates to return to its starting position.

This mundane task is something that we accomplish almost automatically, with very little mental effort, and yet it would defeat nearly all of the most sophisticated present-day robots in a real-life situation. This is a clue that even simple tactile-guided tasks require a rich stream of information (as well as expectations about the physics of our bodies and the outside world). In feeding the parking meter, we rely upon four main types of touch sensor and their associated nerve fibers embedded in the glabrous skin of the fingertips (figure 2.3).

As you begin rooting around in your pants pocket (or perhaps your purse or backpack) trying to identify a quarter by touch alone, you fondle a USB thumb drive, two sticky ibuprofen tablets, a dime, two pennies, and a nickel before you recognize a quarter based upon its size and the texture of its bas-relief faces and ridged edge. In this process, all four main types of touch sensor in the skin of your fingertips are active, but the key sensor that allows you to detect the edges of objects, local curvature, and rough texture is called a Merkel cell. It is named after the German anatomist Friedrich Merkel, who first described them in 1875, calling them Tastzellen, or "touch cells." These specialized epidermal cells are found clustered in disks made up of several cells. The disks are located on the peaks of the primary epidermal ridges, at the border between the epidermis and the dermis (figure 2.3). A Merkel disk is contacted by a single nerve fiber that transmits information from the Merkel disk–nerve fiber junction in the skin to the spinal cord and, ultimately, the touch-sensing regions of the brain. This electrical information is encoded by brief changes in voltage lasting about one one-thousandth of a second, called spikes.[14] One enduring question is how the mechanical energy of skin deformation is transformed into an electrical signal in the nerve ending. The best present hypothesis is that this is accomplished by a type of

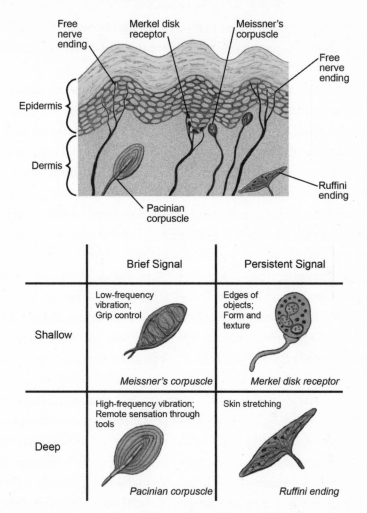

Figure 2.3 Four types of sensor for mechanical stimuli found in glabrous skin. Merkel disks are located in the deepest part of the epidermis, where it borders the dermis at the peaks of the primary epidermal ridges. Meissner's corpuscles are found just across this border in the shallowest parts of the dermis but in the troughs between epidermal ridges, while Pacinian corpuscles and Ruffini endings are located deeper in the dermis. The nerve fibers receiving signals from Meissner's and Pacinian corpuscles send electrical signals to the brain briefly, only at the beginning and end of a sustained touch, while those that receive signals from Ruffini endings and Merkel cells signal persistently, throughout the touch stimulus. Also shown are free nerve endings, which are sensors for certain chemicals, temperature, pain, and itch. Those will be discussed in later chapters.

molecule embedded in the membrane of the nerve ending called a stretch-activated ion channel. These molecules form a pore that is closed at rest, but when the cell membrane is stretched the pore opens, allowing positive ions like sodium and calcium to flow into the nerve cell and thereby trigger an electrical spike.[15]

Merkel disks are present at a very high density in the skin of the lips and the fingertips, at a low density on other glabrous regions, and at a very low density on hairy skin. They are sensitive to very small forces that produce skin indentations of about 0.05 millimeter, and continue to respond more strongly (firing spikes at a higher rate) in a linear fashion, until they fire maximally in response to indentations of about 1.5 millimeters. Electrical recordings made from single nerve fibers conveying Merkel disk signals reveal that these fibers continue to fire spikes as long as the skin remains indented.[16] Artificial electrical stimulation of a single Merkel nerve fiber running through the upper arm causes subjects to report a sensation like a "soft painting brush held tangentially against the skin."[17]

Merkel disks allow us to distinguish individual surface features with our fingertips, like the rough-textured ridges on the edge of the quarter. Crucially, the ability of Merkel disks to distinguish tactile features flows from their particular structure, location, and connections. Because Merkels are located in a relatively shallow layer of the skin, they can respond to small indentations produced by textured surfaces. And because they are densely packed in the fingertips, and each is innervated by a single nerve fiber, this array of sensors can resolve the difference between two features on the surface of an object that are only about 0.7 millimeter apart.[18]

Okay, now that you've identified the quarter, you grip it between your thumb and forefinger and prepare to maneuver it toward the coin slot. How do you determine the amount of force to exert with this pincer motion? You don't want to use maximal bone-crushing force for everything you grip—which might not be bad for holding a quarter, but would be disastrous for clutching an egg or a child's hand—nor do you want to use so little force that the quarter slips out of your grasp. Ideally, you'd like to use the minimal amount of force needed to hold the quarter securely. For this job you rely mostly on another skin sensor, called the Meissner's corpuscle (figure 2.3). Like the Merkel disks, the Meissner's corpuscles are located at the border between the dermis and the epidermis.[19] They reside just on the dermal side of the border, but in the troughs between the ridges, where the epidermis is thinnest. Each Meissner's corpuscle consists of a coiled arrangement of nerve fiber endings, intermingled with layers of nonneuronal cells called Schwann cells. Together these form a bulbous encapsulated structure, the corpuscle, which is tethered to nearby skin cells by structural cables made from the protein collagen.[20] Meissner's corpuscles are physically deformed by the tug of these cables when the skin is indented and pop back into shape when the indenting object is removed.

The array of Meissner's corpuscles in the fingertips is even denser than that of Merkel disks, and they are located even closer to the skin surface. These properties might lead one to believe that Meissner's corpuscles are also built to convey information about fine features of objects, such as texture, edges, and curvature. However, when electrical recordings are made from the nerve fibers that innervate Meissner's corpuscles, we find completely different responses. First, Meissner fibers fire spikes only at the very beginning and the very ending of a prolonged skin indentation: when the outer capsule is initially deformed and then again when it pops back into place. This means that, unlike Merkel disks, Meissner's corpuscles don't respond well to steady force on the skin, but rather are strongly activated by faint low-frequency vibration that repeatedly indents and reforms the capsule. Second, a single nerve fiber conveys and collects signals from many Meissner's corpuscles, spread over about 10 square millimeters of skin surface. Even though Meissner's corpuscles are found in the fingertips at great density, electrical recordings show that they cannot distinguish the finest features of objects. The convergent wiring of the Meissner system is built to be exquisitely sensitive to tiny, rapid skin movements but to localize those movements with only moderate precision.

What does all this have to do with gripping your quarter properly? It turns out that when you grip and move an object, there are microscopic slips of that object along your skin. These microslips are detected by the Meissner system, which sends electrical signals to neurons in the spinal cord that contract the relevant finger muscles to increase gripping force until the microslips stop. This allows you to manipulate objects with delicacy, using the minimal force for the job at hand. Because Meissner-mediated grip control is a spinal cord circuit, it works as a reflex and does not directly enter your conscious awareness—you don't have to think about gripping the quarter slightly harder as you move it from your pocket to the coin slot, as it just happens naturally.

To appreciate how the anatomy and physiology of the Meissner's array in the fingertips make precision grip control possible, let's indulge in a little science fiction. Imagine an alternate human biology in which Meissner's corpuscles signaled throughout a period of skin indentation rather than just at its beginning and end. If this were the case, they would become sensitive to ongoing force, rather than insensitive, as they are in our fingers. In that alternate-skin universe, the Meissner's corpuscle responses to the large steady forces required to grip an object would overwhelm the small signals produced by local microslip vibrations. The useful signal about grip efficiency would drown in a sea of noise, and fine grip control would fail. Without fine grip control, we would have been unable to develop tool use and, very likely, human culture as we know it. Sometimes even the tiniest, obscure-seeming details of biology turn out to be critically important.

By now you're ready to place the quarter in the meter's slot. As you insert the coin, you begin feeling sensations transmitted through it as it collides with the internal walls of the slot and you will use this tactile feedback subconsciously to alter the trajectory of your arm, hand, and fingers in order to insert the quarter smoothly. For this portion of the task, the most important sensor is called the Pacinian corpuscle.[21] Pacinian corpuscles look weirdly lovely (figure 2.3), each consisting of a single nerve fiber ending wrapped in many concentric layers of supporting cells with fluid-filled spaces. In cross-section, they look like an onion or an entry in one of those engineering contests for high school students that involves designing a lightweight casing to protect an egg dropped off the roof of a building.[22] There are about 350 Pacinian corpuscles per finger, located in the deeper regions of the dermis. Electrical recordings from Pacinian nerve fibers reveal that, like Meissner's corpuscles, they don't respond well to continuous force but fire spikes only at the onset and offset of skin indentation. Although they are also poor at resolving the surface features of an object, Pacinians are extremely sensitive to tiny vibrations and have almost no spatial localization: A single Pacinian corpuscle in the fingertip, by virtue of its layered wrapping and deep location, can be activated by vibration occurring anywhere on the finger. In a way, the properties of the Meissner's corpuscle (sensitive to small vibrations but insensitive to steady force or fine spatial detail) are even more extreme in the Pacinian corpuscle. Pacinian corpuscles are most sensitive to high-frequency vibration in the range of 200 to 300 hertz, at which they can detect skin motion as small as 0.00001 millimeter (two hundred times smaller than the diameter of a tiny vellus hair).

When I was a child, I loved to watch the seismograph at the Griffith Park Observatory in my hometown of Los Angeles. With its ink pens tracing wiggling lines on chart paper, this exquisitely sensitive instrument could detect vibrations from an earthquake in Japan, propagated across the entire Pacific Ocean, or from a bomb test in the Nevada desert, hundreds of miles away. But it could also be activated by thirty rowdy schoolchildren jumping up and down together in the same room that housed the instrument. (Believe me, we did the experiment.) Without contextual information provided by other seismographs in different locations, however, the Griffith Park instrument would not be able to tell one event from another. Roughly speaking, the Pacinian corpuscle is built with the same engineering trade-off as the seismograph: extreme sensitivity to vibration at the expense of localization.[23]

Another role of Pacinian touch sensors is to provide a high-fidelity neural image of transient and vibratory stimuli transmitted to the hand by an object held in the hand. This object can be a quarter, as in our example, but, more important, it can be a tool or a probe. When we use a tool, like a shovel, we can perceive tactile events at the working end of the tool almost as if our fingers were present there. Imagine digging into a pile of gravel with a shovel and then doing the same with a pile of soft, loose topsoil. You can easily distinguish the different properties of gravel or topsoil through the shovel, even though your hands are far away from the contact point. Furthermore, with practice, our ability to interpret this kind of long-range touch information improves. In this way, the violinist's bow, the surgeon's scalpel, the mechanic's wrench, or the sculptor's chisel effectively become sensory extensions of the body.

This effect is not limited to simple tools. Automotive enthusiasts become rhapsodic over "road feel," the fidelity of tactile information about the road surface transmitted to the driver's hands through an entire series of linked mechanical parts (tires, wheels, tie rods, steering column, steering wheel). And they get upset when technological changes interfere with road feel at the expense of other features, as in this review of the 2013 Porsche Boxster by Lawrence Ulrich of the *New York Times*:[24]

> Like every other company desperate to increase gas mileage, Porsche is replacing traditional hydraulic steering with electric-assisted units. Describing the difference in feel between hydraulic and electric steering isn't easy.

But traditionally, steering a Porsche was like shutting your eyes and running your hands over a face—every crease, stubble and dip comes through your fingertips, the image of the road becoming clear. The Boxster's electric steering delivers more muted sensations.

So next time you're driving your old-school Porsche with hydraulic steering and appreciating the subtle sensations of road feel, consider how your Pacinian touch sensors are shaping your experience. What's more, even if you've mashed the gas pedal too hard and are now tightly gripping the wheel with fear, you'll still be able to perceive those fine road-feel sensations, because the Pacinian corpuscles report only the high-frequency vibrations transmitted through the steering wheel, not the constant force exerted by your white-knuckled fingers.

Returning to our sidewalk, as you hear your coin drop into the meter, you begin grasping and twisting the meter's handle. This action will activate all three of the previously mentioned touch sensors. The Merkels give you information about the edges and curvature of the handle as well as the steady force of its resistance to your turn. The Meissners give you low-frequency vibration and microslip signals that you use reflexively to fine-tune your grip strength. And the Pacinians transmit the high-frequency vibrations of the meter's internal ratchet mechanism. The fourth system that comes into play appears to be involved in sensing horizontal skin stretching and is called the Ruffini ending. The Ruffini ending forms an elongated capsule in the deep dermis in which the ends of nerve fibers intermingle with collagen fibers of the skin (figure 2.3).[25] The long-axis of the Ruffini ending typically runs parallel to the skin surface, which may explain why they are highly responsive to horizontal stretching and less sensitive to skin indentation. Ruffini endings are present at much lower density in the skin of the hand than the other three sensors, so they have poor spatial resolution. Recordings from Ruffini nerve fibers show that they fire throughout a prolonged stretch stimulus and are only weakly sensitive to vibration. Stimulation of single Ruffini nerve fibers can sometimes give rise to a sensation of skin stretching.

How the brain makes use of the information from Ruffini nerve fibers is poorly understood. Ruffini signals may help detect motion of an object along the skin surface as that object stretches the skin locally. More interesting is the idea that Ruffinis provide information to the brain about the conformation of the hand and fingers through skin stretch signals: For example, as you extend your fingers, the glabrous skin of your fingers and palm is stretched.[26] It has also been suggested that Ruffinis may perform a similar function in other locations where horizontal skin stretch indicates position of a limb. For example, the hairy skin over the elbow is stretched as the elbow joint is flexed, and this may help inform the brain about the status of the arm and its readiness for certain movements.

Looking at the four types of glabrous skin touch sensors in figure 2.3, we see an appealing functional symmetry: Two receptors are shallow, and two are deep; two signal briefly, and two, persistently. All the possibilities are being covered systematically. These four streams of information remain separate as they are conveyed to the spinal cord. A single nerve fiber is dedicated to one type of sensor: It won't contact both a Ruffini ending and a Pacinian corpuscle, for example. Each of these four types of nerve fiber is a "labeled line" built to convey a single type of information onward to the spinal cord and brain stem.[27]

The four touch receptor systems we've explored are called mechanoreceptors because they have the common property of converting mechanical energy delivered to the skin into electrical signals. There are also sensors in the skin that react to nonmechanical stimuli. Both hairy and glabrous skin have free nerve endings that terminate in the epidermis (figures 2.3 and 2.4) and are involved in sensing pain, itches, certain chemicals, inflammation, and temperature. File that information away for now, and we'll return to discuss those other skin senses in later chapters.

~~~~~~

In college I had a friend named Chuck, a dedicated, muscle-bound competitive swimmer who would routinely shave his arms, legs, and chest in the belief that his depilated body would glide through the water more efficiently. I was not convinced that his motivations were entirely hydrodynamic in nature. When I teased him about this, his eyes lit up, and he confided in a low, manly voice, "It may or may not make any difference to my swimming speed, but I dearly love the way it feels when I slip between the sheets at night."

Touch sense in hairy skin has been investigated much less thoroughly than that of glabrous skin. Hairy skin comprises all four of the classical mechanoreceptors found in glabrous skin, although typically at much lower density. As Chuck appreciated, much of the sensation from hairy skin comes from the way the hairs and the surrounding tissue interact. In hairy skin, Merkel cell–nerve fiber complexes can be found in clusters around the base of guard hair follicles, where they can be deformed by bending of the hair, resulting in a persistent signal. However, the main sensory signal from hair deflection is brief and is provided by specialized bare nerve fibers that enclose the base of the hair follicle in a pattern that resembles the vertical bars of a jail cell (figure 2.4). These are called longitudinal lanceolate endings, and they can detect very small hair deflections. Like our pet cats, we know that it does not feel the same to receive a caress in the direction of the angled hairs as it does against the grain. This results from the ability of longitudinal lanceolate endings to respond differently to hair deflection toward the skin versus way from it.[28] Hairs are also innervated by lasso-shaped circumferential endings, and these nerve fibers appear to be particularly sensitive to hair pulling, much to the delight of boys everywhere.[29]

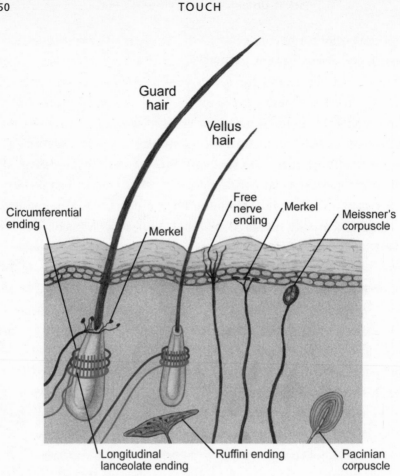

Figure 2.4 Innervation of hairy skin. Guard hairs have clusters of Merkel endings surrounding the outermost portion of the hair follicle. Both guard and vellus hairs are innervated by longitudinal lanceolate and circumferential endings. Here, longitudinal lanceolate endings are shown as a single population. In fact, there are at least three different types of longitudinal lanceolate ending, each of which conveys a slightly different signal in response to hair deflection. When comparing the anatomy of the touch sensors in glabrous and hairy skin, it becomes clear that, while these two types of skin are contiguous and developmentally related, they are essentially two different organs, each of which has evolved to detect different types of tactile stimuli.

Louis Braille, the youngest of four children, was born in 1809 at Coupvray, a small town located about twenty-five miles east of Paris. His father, Simon-René, was a successful leatherworker and, even as a toddler, Louis liked to play in his workshop. When he was three years old, Louis went to the workshop to play with an awl. He rested his head on the workbench to get a close-up view, and as he tried to punch through a scrap of leather, horrifyingly, the sharp tip of the awl skittered over the tough surface of the leather and stabbed him in the eye. The injured eye became infected, and the infection eventually spread to his other eye, leaving him completely blind by age five. (This occurred long before the development of antibiotics.) Louis soon learned to navigate his town by using a cane fashioned by his father, who was determined to have the boy engage fully with the world around him. Louis impressed the teachers in his local school with his intelligence and drive, and so, at the age of ten, he was offered a place at a special boarding school, the newly founded National Institute for Blind Youth in Paris.

This school, one of the first in the world for blind children, was founded by philanthropist and director Valentin Haüy. Its pupils were taught to read using a system devised by Haüy in which Roman letters formed of copper wire were pressed into thick paper to create raised imprints that could then be "read" with the fingers. The Haüy system was useful but quite limited. Distinguishing individual letters required a lot of fingertip scanning, so reading speed was very slow. Because the letter imprints themselves were necessarily large, a single page could contain only a few sentences, and production of books using the Haüy system was time-consuming and expensive. (There were only three such books for the entire school when Louis first arrived.) And of course, the blind children had no way to write with this method, which required a specialized workshop.

Using the few Haüy books available and listening to lectures, Louis learned quickly but dreamed of an alternative system for reading and writing for the blind, one that would be faster and easier to use. In 1821, when he was twelve years old, he heard of a tactile writing system invented by a French Army officer, Captain Jacques Barbier, who developed his "night writing" for stealthy battlefield conditions where it would be dangerous to speak or shine a light lest one draw enemy fire. Barbier's system, a series of raised dots and dashes, was superior to Haüy's letters because it could be read by a trained soldier with a single pass of the fingers. But it was still too slow and cumbersome for reading long passages of text. Inspired by Barbier's night writing, Louis worked to create a more compact and efficient tactile alphabet. After much tinkering with an awl, the same instrument that blinded him years before, Louis settled upon a compact two- by three-row grid of raised dots to create a code in which each letter of the Roman alphabet had a corresponding unique dot pattern. He also devised a grooved slate and stylus so that the blind could easily write on paper. Impressively, by the age of fifteen, he had produced a near-final version of the writing system for the blind that now bears his name.

Hired as a teacher at the Paris school, Louis went on to publish books about his writing system and another raised-dot code he devised for musical notation. Unfortunately, Braille writing was not adopted during his lifetime, either at the school where he taught or anywhere else. Director Haüy, who was sighted, was most interested in promoting his own tactile writing method, which was also easy for sighted people to read visually.

As a result of overwhelming protest by Louis's students, Braille writing was finally adopted at the Paris school two years after his death from tuberculosis at age forty-three. It soon spread throughout the French-speaking world but took much longer to take root elsewhere, notably in the United States, which did not officially adopt it until 1916. Today Braille is a global standard. There are different Braille systems in use across the world, including languages that use non-Roman alphabets (like Greek and Russian) and those that use a pictogram in place of an alphabet (like Chinese), as well as mechanical Braille presses and even Braille computer interfaces.

The average reading speed for expert Braille readers is about 120 words per minute, and the very fastest Braille readers can blaze along at 200 words per minute.[30] This requires extraordinarily fast processing of tactile information: Each Braille character must be recognized within about one-twentieth of a second (fifty milliseconds). When Louis Braille devised his writing system, he had no knowledge of the properties of Pacinian corpuscles or Merkel endings or sensory nerves. He simply used his own tactile experience to carefully set the spacing of the dots—sufficiently far apart so that a dot could not be mistaken for its neighbor, and sufficiently compact so that the complete two- by three-row grid could fit under a single fingertip.

Of the four mechanosensors in the fingertip, which are tuned to encode Braille characters? To address this question, Kenneth Johnson and his colleagues at the Johns Hopkins University School of Medicine took recordings of single nerve fibers while Braille characters were scanned across a subject's fingertip. They then plotted the electrical activity in a grid to form a visual image of the information transmitted by the four different types of nerve fiber (figure 2.5A). This wonderful experiment demonstrated that only the Merkel fibers faithfully represented the pattern of Braille dots. Meissner's fibers produced a blurry image, while the deep sensors (the Pacinians and Ruffinis) failed to encode the Braille dots at all. When this experiment was repeated using scaled-down Haüy-type raised Roman letters, the Merkel fibers were also able to encode them, but the resultant neural image reveals the ambiguity inherent in the Haüy system. Examining figure 2.5B, you can see that the neural responses to certain letters are easily confused: C, G, O, and Q are nearly identical; R looks like H; and P is similar to F. Indeed, when subjects were asked to name raised Roman letters scanned across the fingertip, those groups of letters were the ones that were most often misidentified.[31]

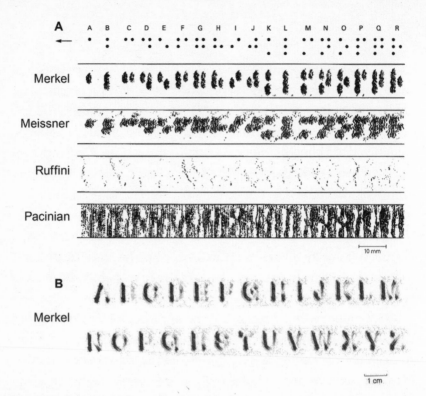

Figure 2.5 The response of single axons innervating the human fingertip to Braille and Haüy-type raised Roman letters. (A) In this example, the Braille characters were scanned across the fingertip at a rate of 60 millimeters per second and electrical activity was recorded in single nerve fibers of various types. The image is created by writing a dot when a spike is fired in the fiber to make a single horizontal line. Then the Braille pattern is moved vertically by 0.2 millimeter and scanned again, and this process is repeated to create the raster image. Only the Merkel responses faithfully represent the Braille characters. The Meissner responses are blurred, and the deep sensors, the Pacinians and Ruffinis, carry no information about the Braille dots. Adapted from J. R. Phillips, R. S. Johansson, and K. O. Johnson, "Representation of braille characters in human nerve fibers," *Experimental Brain Research* 81 (1990): 589–92, with permission from Springer. (B) Merkel responses to raised Roman letters are sufficient to allow for some decoding but are prone to errors of interpretation. From F. Vega-Bermudez, K. O. Johnson, and S. S. Hsiao, "Human tactile pattern recognition: active versus passive touch, velocity effects, and patterns of confusion," *Journal of Neurophysiology* 65 (1991): 531–46, with permission of the American Physiological Society.

I love boozy, geeky conversations in which people drop their inhibitions and consider subjects that might seem silly but actually raise substantive points. Years ago I was talking with my nonscientist friend Q. about tactile sense, and she asked the following wonderful question: If you're a blind Braille reader and you've lost your fingers, could you read with other sensitive skin areas, like the genitals? After all, she said, they do produce a strong sensation in response to light touch. The answer I gave her is that the genitals (both the glabrous and the hairy parts, both male and female) are *sensitive*, in that they can detect tiny skin indentations. However, they are not finely *discriminative*, meaning that they do not allow for determination of the precise location, texture, or form of objects pressed against the skin. These regions fail at fine discriminative touch because they are poorly endowed with the shallow touch sensors, most notably the Merkel disks.

Q. was skeptical of my explanation, and so, after quizzing me on experimental design, she decided to perform her own investigation. She gathered a compass with blunted points, a blindfold, and a willing accomplice (her husband). His job was to repeatedly press the compass points gently into her labia minora, varying the gap from 1 to 20 millimeters, and recording whether she perceived the stimulus as a single point or two separate points. This is a standard technique for mapping the skin and is called a two-point discrimination threshold test. Then the tables were turned, and she did the same on her husband's penis.[32] The kinky results: The two-point discrimination threshold on her labia minora was approximately 7 millimeters. On his penis, it was 5 millimeters on the glabrous skin of the glans and 12 millimeters on the hairy skin of the shaft. This compares to a typical value of 1 millimeter on the fingertip and establishes that crotch-based Braille reading will fail. However, not all erogenous zones have poor discrimination of spatial detail. The lips and tongue, by virtue of having a high density of Merkel endings, have superb spatial resolution and hence can be used to read Braille.[33]

~~~~~~~

The typical neuron has a cell body, which contains the DNA-packed nucleus and other organelles, and two different types of projecting fiber, dendrites and axons. The dendrite, a branched signal-receiving structure, passively conducts electrical signals through the cell body to the axon, the information-sending part of the neuron. There is a specialized zone at the start of the axon that can trigger all-or-none signals, the spikes. These spikes can be propagated in a regenerative fashion (like a flame moving down a fuse, igniting the next segment of fuse continually) over long distances. When the spike reaches the terminal end of the axon, it triggers, through a series of rapid biochemical steps, the release of a chemical neurotransmitter that diffuses across a tiny fluid-filled gap to activate specialized neurotransmitter receptors in the dendrite of the next neuron. This connection between neurons is called the synapse. The process by which electrical signals are converted into chemical signals and then back again into electrical signals in the receiving neuron is called synaptic transmission.

The neurons that convey tactile information from the skin to the spinal cord and brain do not have this typical dendrite–cell body–axon shape. Instead, they have a single long axon running from the spot in the skin they sense to the spinal cord. The cell body is attached to the axon by a little stub, tucked off to the side. The cell bodies of many sensory neurons are clumped together in a structure called the dorsal root ganglion, which lies just outside the spinal cord (figure 2.6). There are many pairs of dorsal root ganglia (one on each side of the body), each pair associated with one of the stacked spinal bones called vertebrae.[34]

When we think of electrical signaling, we imagine impulses in our laptop computers or iPods, which move at a speed slightly slower than the speed of light, about 669 million miles per hour. Transmission of electrical spikes in the nervous system is a much, much slower process. The axons that carry information from your skin mechanoreceptors can convey spikes at about 150 miles per hour (70 meters

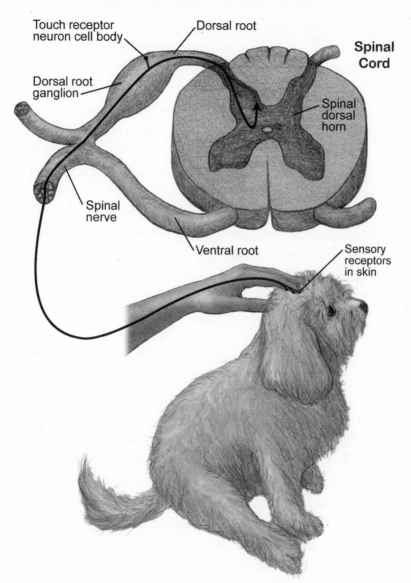

Figure 2.6 Electrical signals from sensors in the skin are conveyed to the spinal cord and from there to the brain (upward arrowhead) along the axons of neurons whose cell bodies reside in the dorsal root ganglia. © 2013 Joan M. K. Tycko

per second). These are some of the faster axons in the nervous system, yet they are still more than four-million-fold slower than signals in electronic devices. Or, in other words, imagine if a giant were lying with her head in Baltimore and her foot dangling in the waters off Cape Town, South Africa. If she were brushed on her big toe by a frond of seaweed on Monday at noon, activating her skin mechanoreceptors, she wouldn't feel it until Wednesday in the midafternoon, when the signal would reach the neocortex of her brain, and she wouldn't be able to twitch her foot in response until early Saturday.[35] Leaving giants aside for a moment, the key point is that it takes time for electrical signals that begin in the skin to propagate to the brain, where they are perceived, and it takes longer for these signals to travel from distant parts of the body, like the toes, than from closer locations, like the face.

<center>~~~~~~</center>

At the turn of the twentieth century, some Europeans and Americans, mostly women, would go to the doctor reporting that they had lost touch sensation in a particular region of the body—only a weird, vague tingling remained. Is there a neurological explanation for these symptoms? We know that nerves to the spinal cord and brain convey tactile signals. So might the problem have been that something was interfering with the functioning of a particular sensory nerve—pinching or an infection, perhaps—and it was this disruption that gave rise to a localized numbness? Figure 2.7 shows an image of the human body drawn so that the skin territory innervated by each pair of dorsal root ganglia (one on the left and one on the right) is mapped upon it. For example, you can see that the fourth thoracic dorsal root ganglia carry sensations from a horizontal band of skin that wraps around the torso at the level of the nipples, while the first sacral ganglia, lower down in the spinal column, innervate vertical stripes running down the outside of the calf, ankle, and foot. Each of these skin regions innervated by a pair of dorsal root ganglia is called a dermatome.

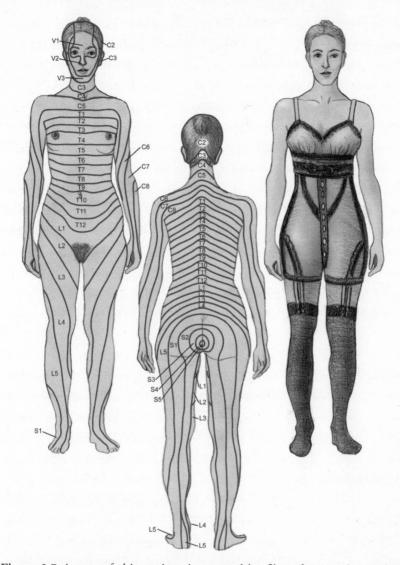

Figure 2.7 A map of skin regions innervated by fibers from various spinal nerves and the trigeminal nerve serving parts of the face. This map (left and center) shows that these dermatomes do not correspond to the underwear-shaped regions of numbness (right) widely reported by female patients at the turn of the twentieth century. The letters S, L, T, and C correspond to groups of vertebrae, ranging from the bottom to the top of the spinal cord: sacral, lumbar, thoracic, and cervical. The letter V denotes the trigeminal nerve, which originates from the brain stem. It is the fifth cranial nerve, hence it is designated by the Roman V. © 2013 Joan M. K. Tycko

If trauma to a sensory nerve or a dorsal root ganglion were the cause of these tingly symptoms, one would imagine that the reported area of numbness would have the shape of a dermatome (or maybe two adjacent dermatomes). However, when the doctors of that era systemically probed the skin to map out the area of lost sensation, a different story emerged. Most often the region of numbness corresponded not to the configuration of a dermatome but rather to the shape of various types of underwear of the period: corsets, drawers, bloomers, garters, stockings, and so on. This led many doctors of the time, including Sigmund Freud, to conclude that these numbing symptoms did not have a basis in damage to the sensory nerves, but that the perception of underwear-shaped numbness originated in the brain as a result of psychological and social factors. Today, underwear-shaped numbness is a much less common medical complaint. (It's harder to imagine a numb thong-shaped region.)

~~~~~~

We think of neurons as being microscopic in scale, like other cells, and in one sense this is true: The cell bodies of sensory neurons in the dorsal root ganglia range from 0.01 to 0.05 millimeter in diameter. (The very largest of these cell bodies is approximately the same diameter as a human guard hair.) However, it's remarkable how long certain sensory neurons must be to convey touch signals. Let's consider the axon of a mechanosensory neuron that innervates the heel. It runs from the heel up the leg into the pelvis and through the dorsal root ganglion to enter the spinal cord at the first sacral spinal nerve. It will then continue to ascend all the way up the spinal cord, finally terminating and forming synapses in a region of the brain stem called the gracile nucleus. In a typical person that neuron is about 5 feet long. (And, yes, in a giraffe it's longer still.) These neurons are the longest cells of any type in the body.

The gracile nucleus is not the endpoint for fine touch signals, however, but merely the first stop along the way. The axons of gracile nucleus neurons ascend farther, cross over to the opposite side of the brain, and convey their electrical signals onto another processing station in a region called the thalamus, which in turn sends its axons to the cortex, the huge rindlike covering that forms the surface of the brain.[36] The region where these axons from the thalamus terminate is called the primary somatosensory cortex ("primary" because it's the first of several regions in the cortex that receive touch information). This region is located in a strip just behind the central sulcus, the main fissure that divides the brain into front and back portions (figure 2.8, left center panel). Because the axons carrying touch information cross the midline of the body before they reach the cortex, the right side of the cortex responds to touch information from the left side of the body and vice versa.[37]

~~~~~~

In the late 1930s, Wilder Penfield, Herbert Jasper, and their colleagues at the Montréal Neurological Institute began using electrodes to locally stimulate the brains of epileptics during surgery. They did this in order to identify the precise region of the brain that triggered their patients' seizures, the so-called epileptic focus, so that it could be removed while causing the least damage to healthy neighboring tissue. There is no one spot in the brain that is consistently the origin of epilepsy, so mapping the epileptic focus had to be done individually for each patient. In this remarkable procedure, a patient's head was shaved and stabilized and the scalp was incised and retracted. Then Penfield used a miniature saw to cut away a circular flap of bone, about the diameter of a tennis ball, which was removed and put aside for later reattachment. Because brain tissue has no pain or mechanosensors, this procedure could be performed under local anesthesia to numb the scalp, underlying bone, and brain-wrapping membranes, while leaving the patient entirely conscious. Penfield wielded a handheld stimulating electrode: a device about the size and shape of an electric toothbrush with a metal needle on the business end and a wire attached to the opposite side that was connected to a device that could deliver weak electric shocks to artificially activate the neurons at the electrode tip.

Penfield would methodically move the electrode across the exposed surface of the brain, asking the patient, "What do you feel now?" at every location. The patient might respond, "I feel a tingling in my left wrist," or, "I smell burnt toast," or, "I'm hearing a bit of music that I last heard during childhood." When stimulating just in front of the central sulcus, simple movements were evoked: a foot jerked, a fist was clenched, or the tongue was extended. Stimulation of the primary somatosensory cortex gave rise to buzzing or tingling sensations at various places on the opposite side of the body. The patients reported that these brain-evoked sensations didn't feel like natural touches and would never be mistaken for normal sensory experience. Rather, they were like crude simulacra of touch experience, lacking some essential richness or context. Penfield had an assistant record every movement or report from the patient in a notebook with numbered lines. Then, after stimulating, he would insert a pin with a tiny numbered flag attached to match the stimulation site with the logged response. After a while, the surface of the brain looked like a weird miniature version of putt-putt golf, with the ridges and grooves of the brain as hazards instead of windmills and ramps.[38] When the flags were photographed and the responses over the extent of the primary somatosensory cortex were analyzed, a wonderful pattern was revealed: A map of the body surface exists in the primary somatosensory cortex. When the touch signals from the skin come to the brain stem and are then passed onward to the thalamus and the cortex, they aren't completely jumbled up. Rather, axons that innervate adjacent patches of skin remain near each other and, with some notable exceptions, those neighbor relationships are preserved all the way up to the cortex to form the touch map.[39]

However, the representation of the body on the cortical touch map is a bit weird (figure 2.8, left center panel), as its constituent parts have been chopped up and reassembled, so that the forehead adjoins the thumb, for example, and the genitals, both male and female, are adjacent to the toes. (We'll revisit the genital map in greater detail in chapter 4.) Also, certain parts of the body are hugely magnified in the map: the hands, lips, and tongue are enormous. The feet are only somewhat enlarged, while the legs, back, torso, and genitals are comparatively tiny. Of course, the configuration is clear: The regions that are magnified in the cortical map are those that have a high density of mechanoreceptors in the skin, particularly the Merkels, which subserve fine discriminative touch.

Is the magnification of certain skin areas in the touch map unique to primates, with their supersensitive fingers and lips? To address this question we can examine animals that perform fine discriminative touch with other structures. One of the most striking examples is a semiaquatic North American mammal called the star-nosed mole (figure 2.8, top right panel). This small, nearly blind creature, which is about twice the size of a mouse, digs tunnels with it strong forepaws near streams and ponds, where it can both burrow and swim. Star-nosed moles tend to evoke extreme reactions: Some people (like me) find them cute, but to others their appearance is deeply disturbing.[40] Eleven pairs of fleshy appendages, called rays, surround the nose of this animal. Each ray is endowed with specialized clusters of Merkel disks, together with Pacinian corpuscles and free nerve endings, to form exquisitely sensitive tactile organs, perhaps the most sensitive in the mammalian world. Star-nosed moles deploy their rays in constant motion, exploring between ten and fifteen locations every second. When they contact prey in the form of a worm, snail, or small fish, they devour it immediately; the length of time from touch to ingestion is about 120 milliseconds. Not surprisingly the touch map in the primary somatosensory cortex of the star-nosed mole shows huge magnification of the star at the expense of the trunk, tail, and hind paws (figure 2.8, center

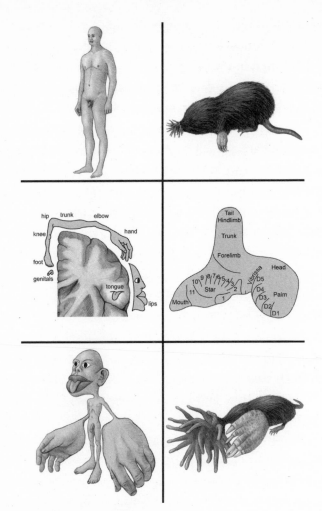

Figure 2.8 Touch maps in the primary somatosensory cortex of the human and the star-nosed mole. Left, top: an adult human; middle: the touch map in the human somatosensory cortex; bottom: a drawing of a human with each body part scaled to the size of its representation in the touch map. Note the huge hands, lips, and tongue. Right, top: an adult star-nosed mole; middle: the touch map of the star-nosed mole in the primary somatosensory cortex (shown in a different plane than the human); bottom: a drawing of a star-nosed mole with each body part scaled to its size in the touch map. Note the expansion of the star and forepaws. Illustration by Joan M. K. Tycko. The left top and middle panels are reprinted from David J. Linden, *The Accidental Mind: How Brain Evolution Has Given Us Love, Memory, Dreams and God* (Cambridge, MA: Harvard/Belknap Press, 2007), 85, with permission of the publisher.

and lower right panels). This design principle holds true across many species: Skin regions with high densities of mechanosensory receptors are magnified in the brain's primary touch map.[41]

When you dig a little deeper, you discover that there is, in fact, a series of touch maps located in adjacent regions of the cortex. The primary somatosensory cortex in primates can be broken into four different smaller areas, each with its own weirdly distorted map. In addition to receiving information directly from the thalamus, these four areas are highly interconnected with one another. They also send information onward to a set of adjacent areas collectively called the higher somatosensory cortex, which does not receive direct contacts from the thalamus. At present, a total of ten different maps of the body have been found in the somatosensory cortices of the primate brain (four in the primary cortex, six in the higher cortex), and more may yet be revealed.

So where in all of these many interconnected brain maps does the miracle happen? How does our rich, nuanced, and profound sense of touch emerge from this crazily wired system of thinking meat? We consciously perceive that a particular location on the skin has been touched because a particular group of neurons in the cortex is activated. This can occur naturally, through stimulation of the skin, or artificially, through direct brain stimulation, as performed by Penfield. But this is only the first small part of an explanation. The truth is that much remains to be understood about how the brain creates our experience of the tactile world. I will not discuss here all the complex connections within and between these various touch-processing cells and regions of the cortex, but there are a few general principles that are worth knowing.

Information converges from skin to brain. Each neuron in the primary somatosensory cortex ultimately receives converging input from many nerve fibers innervating adjacent regions of the skin. For example, a single nerve fiber running through the arm, conveying Merkel disk information from the fingertip to the spinal cord, will respond to stimulation only in a small spot on the fingertip, about 1 millimeter in diameter.[42] But because of this convergent signaling, a Merkel-driven neuron in the finger map of the primary somatosensory cortex is likely to respond to stimulation over a much broader area, about 5 millimeters in diameter. Importantly, this convergent signaling isn't random and thereby doesn't merely degrade the resolution of tactile information. Rather, by wiring a particular group of Merkel-driven fibers to the same cortical neuron, one can construct a cortical neuron that responds to a particular touch feature, like a thin bar laid across the finger pad at a particular orientation. Convergent signaling is even more dramatic in areas that are sparsely innervated: A single neuron in the portion of the touch map representing the back can be activated by stimuli ranging over 50 square centimeters of skin, an area about the size of a playing card.

In the primary somatosensory cortex, convergence produces minimal blending of information: Signals from the four mechanosensory receptors are kept largely, but not entirely, separate. Some column-shaped groups of neurons respond preferentially to Merkels, while others respond to Meissners and yet others to Pacinians. So, for example, a single column of cortical tissue, about 0.6 millimeter in diameter, might receive Meissner signals from the pad of the left big toe, while another receives Merkel signals from the right side of the lower lip.[43]

Serial processing in the brain extracts increasingly complex touch information. If we examine a block diagram of the areas in the cortex that process touch information, it looks at first glance like a random mess, a bowl of spaghetti (figure 2.9). Upon further examination, however, some themes begin to emerge. There are four areas that comprise the primary somatosensory cortex, all of which receive axons directly from the thalamus, though area 3b receives the lion's share. By contrast, area 2 gets a bit of information directly from the thalamus but is also activated by all three other primary areas in the neocortex: 3a, 3b, and 1.

If we make recordings of the neurons in area 3b while stimulating

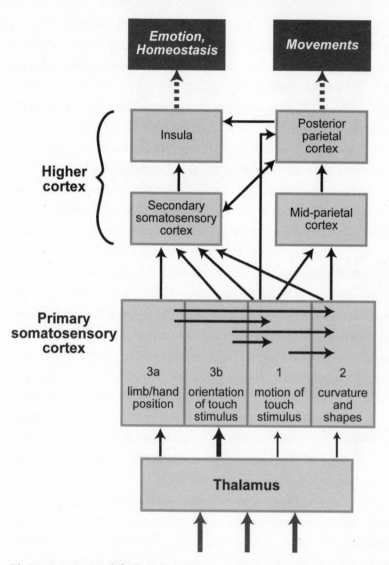

Figure 2.9 A simplified wiring diagram of touch-processing regions in the brain. The primary somatosensory cortex receives information directly from the thalamus, performs mostly serial processing to extract different tactile features, and passes the information on to higher somatosensory cortices. The secondary somatosensory cortex is crucial for the recognition of objects and then conveys this information to brain regions involved in emotion and homeostatsis (the insula) and in the early planning of movements (the posterior parietal cortex). Adapted with permission of Professor Steven Hsiao (Johns Hopkins University School of Medicine).

the skin, we find that the most effective stimulus to activate them is rather simple, like a thin bar laid across the fingertip at a particular angle. This same simple stimulus will only weakly activate neurons in area 2, however, which respond strongly only to more complex stimuli, such as two- or three-dimensional shapes (like a baseball grasped in the hand). When laboratory animals sustain damage to area 3b, the result is remarkable: The animals become touch-blind, to the point that they seem to be almost entirely unaware of the quality or even the existence of touch stimuli. By contrast, damage to area 2 produces a much more subtle effect: These animals can still detect texture, but their ability to recognize objects by touch alone is lost.

The brain's wiring diagram for touch-processing regions enables us to make sense of these findings. Area 3b receives basic touch information that is minimally processed. As a result, neurons in this region respond to simple stimuli, as one would predict from a straightforward convergence of axons conveying mechanosensory signals. Because area 3b is a bottleneck for touch information, damage to it is devastating, depriving most of the brain's downstream processing stations of crucial data. Area 2 has the benefit of receiving not only direct signals from the thalamus but also information from the other primary touch areas, reflecting the processing and computations that those areas perform. As a consequence, area 2 can extract more complex features of touch stimuli, such as object motion, object curvature, and three-dimensional form. When area 2 is damaged, the effects on touch sensation are more subtle, because it is only one of several pathways for information flow to higher areas.

This theme of serial processing with increasing complexity continues as touch information flows farther to the secondary somatosensory cortex. Neurons in this region integrate signals over larger areas (like an entire hand or foot), including regions on both sides of the body. The secondary somatosensory cortex plays an important role in object identification, particularly through exploratory touch. Lesions in this area produce subtle impairments, like loss of the ability to learn a complex object recognition task with one hand and then perform it with the other hand.[44]

Serial complexity appears to be a common theme in the sensory portions of the brain. The visual system, for example, builds recognition of complex visual objects, like faces, from initially primitive features, like spots and bars.

Parallel processing segregates complex touch information into different streams for action. Ultimately, representing the tactile world in the brain is in service of achieving some particular outcome: making a decision, forming a memory, or initiating an action. Heavily processed information flowing out of the higher somatosensory cortex segregates into two different streams. One stream, coursing through a brain region called the insula, informs emotional responses, homeostasis, and some other functions. The insula is now known to be critical for the perceived sense of self. The other stream flows through a region called the posterior parietal cortex and is primarily involved in integrating touch data with information from other senses in order to plan, execute, and fine-tune movements, including the manipulation of objects.[45]

While the primary somatosensory cortex mostly responds reliably and stereotypically to touch information, the brain's higher touch centers are more strongly influenced by cognitive factors like attention, context, motivation, and expectations. We'll return to these regions in later chapters when we consider the higher cognitive aspects of touch.

~~~~~~~

I've discussed how the distorted nature of the human brain's touch maps, with larger areas representing body parts like fingers, lips, and feet, reflects the density of touch receptors in particular regions of the skin. But there's another important factor involved: Touch maps are not permanently defined for one's entire adult life, but can be changed as a result of individual sensory experience. One

nice example of this is found in serious violin, viola, and cello players, professional or semiprofessional, who play for at least twelve
hours per week. For these instruments the digits of the left hand
are continuously engaged in fingering the strings and producing
vibrato effects, a task that involves both enhanced tactile stimulation and exquisite manual dexterity. The right hand, used for bowing,
requires much less individual finger movement and tactile feedback. When string players were placed in brain scanners to measure the representation of their hands in the primary somatosensory
cortex, it was found that the touch map for the left-hand digits was
significantly larger than for the right-hand digits, by about 1.8-fold.
(The right- and left-hand touch maps of age-matched nonmusician
control subjects, meanwhile, were similar in size.) These same basic
results have now been found in three studies performed in different laboratories using somewhat different methods, so the phenomenon appears robust.[46] The interpretation of them, however, is
less straightforward. The simplest explanation is that years of violin or cello practice have produced an enlargement of the left hand's
territory in the touch map. Another suggestion is that people who
happen to be born with or develop unusually large left-hand representations early in life are more likely to take up string instruments
and succeed as players. In the same way that children tend to grav-
itate to sports for which they have a particular natural ability, perhaps they likewise choose an instrument in part based upon some
sense of their innate sensory-motor abilities.

To test these two explanations, you'd need to measure the touch
maps of the hands before and after musical training. Of course,
becoming an accomplished string player takes years, which makes
it difficult to perform such a before-and-after type of study. Are
there tactile experiences that produce changes in the touch map
more quickly? Moving out of the human realm, one striking example is the effect of nursing on first-time mother rats. Norway
rats in the lab give birth to litters of eight to twelve pups. In the
first few days postpartum, they spend about 80 percent of their

time nursing with their twelve nipples, distributed in two rows on their ventral (lower) body surface. When measured twelve to nineteen days after birth, the ventrum portion of the primary touch map of a lactating mother rat was about 1.6-fold larger than that of control rats. (Control rats were either age-matched virgin female rats or nonlactating first-time mothers who had their pups removed at birth.) Fifteen to thirty days after weaning the pups, the ventrum touch map on the lactating mothers had shrunk back to prepregnancy size. These findings suggest that enhanced sensory experience can indeed produce dynamic changes in the touch map, which, at least in some situations, can occur in a matter of days rather than years.[47]

~~~~~

As Tom Waits said, "The large print giveth and the small print taketh away." Experience-dependent plasticity of the touch map works in both directions: Not only can enhanced tactile stimulation cause map expansion, but reduced tactile experience can result in map contraction. When adult rats were fitted with a tiny cast to immobilize one foreleg, the area of the deprived forepaw in the primary touch map was found to shrink by about 50 percent after only seven days; the map representation of the spared forepaw didn't change at all. The authors of this study removed the cast seven days later and checked the touch map again. At that point the deprived forepaw's representation in the touch map was still shrunken. It's likely that, given more time, it would have slowly grown back to its normal size, but the experiment to verify that wasn't done.[48]

All of us are engaged in a slow touch-deprivation experiment throughout our adult lives. From about age twenty to eighty, the density of Merkel disks and Meissner's corpuscles gradually decreases by about threefold (figure 2.10), and fine spatial acuity falls off to approximately the same degree. Does this mean that decreased fine touch sensation in the elderly can be entirely explained by a loss of shallow mechanoreceptors in the skin? Probably not.

One clue is that the reduction in spatial acuity is not precisely uniform over the surface of the body: There's about a 2.5-fold reduction in fingertip sensitivity but about a fourfold reduction in sensitivity on the soles of the feet and the toes.[49] An explanation for this difference is that aging is also associated with a reduction in the speed of spike propagation, from about 150 miles per hour to about 110 miles per hour, in the nerve fibers that convey Merkel and Meissner signals to the brain. This slowing of neural impulses could degrade touch information from distant regions like the toes to a greater degree than it would in closer regions, like the hands or lips. Impaired tactile sensation on the soles of the feet and the toes is an important contributing factor to reduced standing and walking stability in the elderly, often resulting in catastrophic falls.

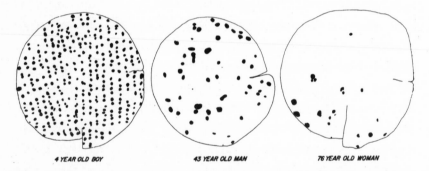

4 YEAR OLD BOY   43 YEAR OLD MAN   76 YEAR OLD WOMAN

Figure 2.10 The density of Meissner's corpuscles decreases with age in the glabrous skin on the plantar surface of the big toe. These drawings are made from 3-millimeter-diameter punch biopsies. In these samples, the density dropped from 47 to 7 to 3 per square millimeter as age increased. From C. F. Bolton, R. K. Winkelmann, and P. J. Dyck, "A quantitative study of Meissner's corpuscles in man," *Neurology* 16 (1966): 1–9, with permission of Wolters Kluwer Health.

Of course the brain doesn't just sit there while we grow older. Experience-driven plasticity in the brain may decline a bit with aging, but it never goes away. Our life experience never stops changing our brains. How the primary and higher somatosensory cortices change and adapt in the face of gradually declining mechanosensor density is not presently understood. And we should not assume that if there are plastic changes with aging, they are necessarily beneficial. Cortical plasticity in response to changing touch information might make the problem worse: In subtly rewiring the circuit incorrectly, it might further degrade tactile acuity.

~~~~~

Ah, the gentle touch of a woman—so subtle, so nuanced, so much better than the ham-fisted groping of a man. Is this the case because her fingertips are more sensitive to the finest details of tactile form? Findings from two different labs initially suggested that this was true. In experiments in which adult subjects were asked to resolve a grooved surface carefully pressed against a stationary fingertip, women significantly outperformed men: On average, they could distinguish grooves that were about 0.2 millimeter closer together.[50] Was this because female fingertip skin is softer and more easily indented? No—women's fingertip skin was determined to be deformed by the grooved surfaces to the same degree as men's. Was it because of some sex-based difference in somatosensory brain circuitry or ability to focus on the task? Perhaps, but there's no evidence to support or disprove that theory.

Daniel Goldreich and his colleagues at McMaster University in Ontario, Canada, had a simpler hypothesis: Maybe, on average, women are better at tactile discrimination because they have smaller fingers. If the same number of Merkel disks, the sensors that subserve the finest discriminative touch, were distributed evenly over large and small fingertips, then the small fingers would have a higher sensor density and therefore greater acuity—it would be like having a 10-megapixel camera in your cell phone instead of a 5-megapixel one. To test this notion they recruited one hundred undergraduates, fifty men and fifty women, and had them perform the grooved-surface discrimination task as a measure of tactile acuity; they also carefully measured the area of the pad of each subject's index finger,

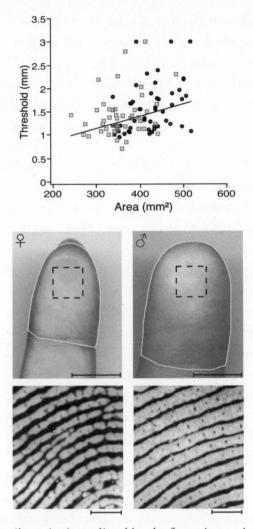

Figure 2.11 Tactile acuity is predicted by the fingertip area in both men and women. Top: a scatterplot relating fingertip area to tactile acuity showing that smaller fingertips are more discriminative. Each plot symbol is a subject. The square symbols are women and the round ones are men. Bottom: High-resolution scans of small female (left) and large male (right) fingertips show that the density of sweat pores, and therefore, presumably, Merkel disk clusters, is higher in smaller fingers. Scale bars = 1 centimeter (top) and 1 millimeter (bottom). This figure is from R. M. Peters, E. Hackeman, and D. Goldreich, "Diminutive digits discern delicate details: fingertip size and the sex difference in tactile spatial acuity," *Journal of Neuroscience* 29 (2009): 15756–61, with permission of the Society for Neuroscience.

which was the one used in the task. Confirming previous results, on average the women bested the men by about 0.2 millimeter. When they made a scatterplot comparing tactile acuity to fingertip area, however, fingertip area proved to be an excellent predictor of fine touch discrimination, a finding that held true for both men and women. Or, stated another way, a man and a woman with fingertips of equal size will, on average, have the same tactile acuity (figure 2.11).

There's no way to directly measure Merkel disk density in fingertips without a painful skin biopsy. But because Merkel disks have been found to cluster around the bases of sweat pores in glabrous skin, sweat-pore density, which can be gauged by wiping the fingertip with water-soluble finger paint and then pressing it against a standard optical scanner, was used as a proxy measure of Merkel disk density. Indeed, sweat-pore density was shown to be significantly greater in smaller fingers. Goldreich and coworkers concluded that finger size predicts tactile acuity independent of sex and that this difference is subserved by increased Merkel disk density in small fingers.[51] Their appealingly compact and parsimonious explanation leads to all kinds of other questions: What about the complement of mechanosensory receptors on other body parts that vary in size? Are there also a fixed number of mechanosensors per leg or breast or penis?[52]

~~~~~~~~

By now you have come to understand the basic idea that there are different sensors in the skin, each tuned to extract a different aspect of information about the tactile world. These streams of information are sent to the brain, where, through a series of serial and parallel pathways, the simple information from individual touch sensors is pooled and combined to extract more complex features of touch, like three-dimensional shape and fine texture and remote sensation at the ends of tools. But don't conclude that the sensors we've considered so far, the fast mechanical touch sensors of glabrous skin, are the whole story. As we'll see, they are only one portion of the full spectrum of touch experience.

CHAPTER THREE

THE ANATOMY OF A CARESS

Baltimore, 1996

I t was hot and stuffy in the jury room. Dust motes hung in the air, and the plumbing pipes ticked. There was a fusty olfactory tableau of cologne, armpit, and lingering cigarette smoke on the clothes of the jurors. The proceedings had been mysteriously halted, and we'd been waiting for hours for the trial to resume. I'd finished my magazine and was left to ponder the testimony thus far. The tale that emerged had raised one of the deep enduring mysteries of tactile neurobiology: What is it, exactly, that makes for a lousy handjob?

~~~~~~

One of the true joys of living in Baltimore is serving on jury duty for the Circuit Court for Baltimore City, an all-too-frequent experience. Each year the dreaded government-green envelope appears, and each year I ready myself for what's usually a single boring day in the courthouse. Although I rarely get picked for a jury, on one of the occasions I did, the defendant, a short, muscle-bound nineteen-year-old night watchman, decided that he would represent himself in court and thereby save the expense and trouble of an attorney. He didn't know that he could challenge prospective jurors, and so I became juror number four.

The prosecutor laid out the case: The defendant's pretty and skittish sixteen-year-old girlfriend had turned up at an emergency room, starved, bruised, and dehydrated. She was reluctant to tell her story, but it gradually emerged. One afternoon she and her boyfriend, the defendant, had been lying in bed at his place, fooling around. He asked for manual stimulation, and she dutifully complied. When he complained that she wasn't doing it right, she tried to vary her technique and add some dirty talk. These flourishes were insufficient, though, and, as he explained, "She would either stroke it too slow or too fast." He rapidly became enraged and then utterly lost control, punching her in the face and chest repeatedly. He then handcuffed her to the bed frame and kept her captive for two days, raping her intermittently. Her testimony was consistent with the medical examination, but she was concerned about the outcome of the case, as she said she still loved him.

When it came time for the defendant to mount his defense, he wheeled a television and VCR into the courtroom and popped a cassette inside. The video showed a house party several weeks after the incident. Teenagers were drinking beer and smoking weed while rap music was pounding in the background. The camera soon found his girlfriend, who, obviously intoxicated, was slurring her words and staggering. Her eyes were bright, wet, and unfocused.

The defendant now asked, "So, that night, I didn't hit you, right?"

"I'm having the time of my life!" she agreed.

"I mean," he asked, "I never tied you up or attacked you, right?"

"What are you talking about?" she countered.

"When you said that I raped you, that was a lie, wasn't it?"

"Whatever," she said. "Put that thing down and party with us."

When the verdict came, the defendant was aghast, unable to believe that the video hadn't saved him. We, the jury, had convicted him on all counts: rape, battery, and unlawful imprisonment. In his statement to the court before sentencing, he tried, none too successfully, to sound contrite: "I'm sorry for what I did. I do have a temper. I'll admit it. And believe me—it was a really bad handjob."

<hr>

Now let's imagine, in the style of a Quentin Tarantino revenge fantasy, that the girlfriend, newly empowered and burning with righteous fervor, has dragged her batterer/rapist into the alley behind the courthouse and forced him to undergo a painful biopsy of the sural nerve. This nerve runs down the back of the calf muscle and innervates the outer edge of the foot. Cutting the nerve and turning it to examine the cut end with a microscope, we'd see a cross-section of different kinds of sensory nerve axons, all intermingled (figure 3.1).[1] The large-diameter axons, called A-fibers, are wrapped in layers of insulating myelin protein to speed neural signals and have several subsets with different functions. One group of A-fibers, called A-alpha, carries very fast information from special sensors embedded in muscles, joints, and tendons. These signals enable you to form a mental image of where your body is in space. This ability, called proprioception, makes it possible for you to sense, for example, the position and the movement of your arm, even when your eyes are closed and you are not touching anything. Another A-fiber, A-beta, conveys fast signals from the skin mechanoreceptors we discussed in chapter 2—the Merkels, Meissners, Ruffinis, and Pacinians that allow for fine tactile discrimination as well as the fast signals from hair deflection. The third type, A-delta, is smaller in diameter and has fewer wrappings of myelin protein, so it conducts signals at medium speed. Some A-delta fibers carry certain aspects of pain and temperature sensation, such as sharp, pricking pain and noxious heat and cold—but more on that later.

The sural nerve also contains much smaller-diameter axons, called C-fibers, which lack insulating myelin. Because of these structural features, electrical signals in C-fibers travel at a leisurely, sidewalk-strolling speed of about 2 miles per hour. By comparison, information from the skin mechanosensors flows at about 150 miles per hour via A-beta fibers, and proprioceptive signals race along at about 250 miles per hour on A-alpha fibers.[2] The conduction speeds of these various fibers constrain the kind of information that they can carry. Fast fibers are necessary to transmit rapidly changing, highly nuanced signals about object shape, texture, vibration, and remote sensing with tools—the kind of fine tactile information encoded by mechanosensors that enable us to discriminate between subtly different touch experiences. For example, Braille writing could not be read if the information it contained was conveyed over slow C-fibers—the fast A-fibers are required for this.

C-fibers, in contrast, are not built to inform the parts of the brain involved in discriminative, factual aspects of touch sensation, but rather function to integrate information slowly and to discern the emotional tone of the particular touch involved—the vibe, if you will. For many years it was thought that C-fibers carried only information about pain, temperature, and inflammation (not the sharp, pricking, well-localized component of pain, but rather its

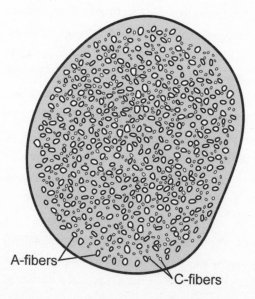

A-fibers

C-fibers

Figure 3.1 Cross-section of the sural nerve. Normal human nerve, like that presumably present in our rapist, showing large myelinated A-fibers and small unmyelinated C-fibers intermixed. This is one fascicle (a bundle of fibers that run together as a cable) in the sural nerve, which typically contains several fascicles.

slow, burning, throbbing, aching aspect, which is so emotionally tax-
ing). However, more recently it has become clear that some C-
fibers convey a special kind of tactile information: They appear to
be tuned for interpersonal touch. These axons, called C-tactile fibers, are caress sensors.

C-tactile fibers innervate only hairy skin, and their endings
wrap around hair follicles, enabling them to respond to hair deflection. We don't yet have unambiguous images of C-tactile fibers in
human skin. However, using genetic tricks in mice, David Ginty
and his colleagues at the Johns Hopkins University School of
Medicine were able to introduce fluorescent molecules into different populations of sensory neurons. This study showed that certain types of hair follicles (giving rise to hairs called zigzag and
awl/auchene types) are innervated by C-tactile fibers. These appear
to be the equivalent of human vellus hairs. Interestingly, these
types of follicles are also innervated by A-delta and A-beta fibers,
and their longitudinal lanceolate endings are interleaved with one
another in a beautiful pattern that looks like overlapping picket
fences (figure 2.4). Obviously, the furry skin of a mouse has a
somewhat different structure from the merely lightly haired skin
of a human. Nonetheless, these findings indicate that hair deflection can give rise to multiple sensations even when a single type of
hair is involved: a fast, discriminative, emotionally neutral signal
mediated by A-beta fibers, and a slow, diffuse, pleasant signal from
the C-tactile fibers.[3]

Studying the role of C-tactile fibers in touch sensation is complicated by the fact that hairy skin is also innervated by fast A-beta
and medium-speed A-delta fibers, relaying signals from hair deflection, as well as by the four conventional mechanoreceptors. It is
not possible to simply stroke hairy skin and measure the percept
evoked by C-tactile fibers alone, either with behavioral tests or a
brain scanner, as any stroke would activate A-fiber responses as
well. For years this problem of overlapping signals impeded our
understating of the caress sensors.

At the age of thirty-two, G.L. became touch-blind. If you ask her, she'll tell you that, in her daily life, she can't feel anything below her nose, and if she closes her eyes, she has no idea where her limbs are in space. G.L.'s neurological deficit is remarkably specific. She is intelligent and does not have obvious problems with cognition or mood. Her ability to contract her muscles and thereby move her body is also intact. However, because she has no proprioceptive sense, she must rely mostly on vision to know where her limbs are positioned. As a consequence, her movements are slow and poorly coordinated, and she must use a wheelchair to ambulate. After extensive physical therapy she has been able to live independently in her home in Québec, Canada, for many years (she is sixty-five at the time of this writing).

A biopsy of G.L.'s sural nerve revealed the origin of her touch-blindness. She has lost her large myelinated fibers: the A-alpha fibers that carry proprioceptive information and the A-beta fibers that convey signals from the skin mechanoreceptors (figure 3.2).

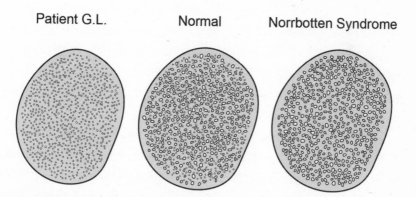

Patient G.L.   Normal   Norrbotten Syndrome

Figure 3.2 A sural nerve biopsy of patient G.L. In G.L.'s sural nerve, note the lack of large myelinated A-fibers and the preservation of small unmyelinated C-fibers. In the center, a normal nerve is shown for comparison. A nerve from a patient suffering from Norrbotten syndrome (also called HSAN V) shows the converse of G.L.'s condition, with spared large A-fibers and decimated A-delta and C-fibers.

Because her A-delta fibers and C-fibers remain intact, her sensations of pain and temperature are normal. She is one of a small number of patients worldwide who have this syndrome, known as acute sensory neuronopathy.[4] Some people with G.L.'s syndrome report that their bodies have come to feel like a foreign entity.[5] They inhabit them, yet they are not entirely theirs. Others have a sense of becoming hyperconscious of their bodies, perhaps because they are focusing their attention strongly to detect the diminished touch sensations that do remain.

G.L. herself has been very generous with her time and has consented to be the subject of many investigations. Although she claims to be entirely touch-blind in everyday life, an interesting exception is revealed in the lab. When a stroke with a soft brush or a gentle fingertip caress is applied to the hairy skin of her forearm and she is asked to concentrate, she has a vague pleasant sensation, with no associated feeling of pain, temperature, itch, or tickle. When attending closely, she can usually tell which arm is being touched but cannot determine the location precisely. Crucially, when these gentle strokes are repeated on the glabrous skin of the palm, she has no sensation at all. These diffuse pleasant sensations are conveyed by her surviving C-tactile fibers, which innervate hairy but not glabrous skin. Amazingly, G.L. and patients like her lack fast, information-rich, emotionally neutral discriminative touch, but seem to retain a dedicated slowly functioning system for diffuse pleasant touch that conveys a feeling of safety.[6]

But are G.L.'s C-tactile fibers functioning like those in uninjured people, or have their properties changed in response to the loss of neighboring A-fibers? The former appears to be the case. When electrical recordings were made from single nerve fibers running in the arm of normal subjects, experimenters could locate C-tactile fibers that responded to a caress of the hairy skin but not to simple skin indentation or vibration. (C-tactile fibers are never activated by any form of touch on the glabrous skin of the palm.)

Single A-beta fibers can be recorded in normal subjects that correspond to the various types of skin mechanosensors described earlier (Merkel, Pacinian, etc.), but they all have response properties that distinguish them from C-tactile fibers. A-beta fibers will respond to both a forearm caress and other forms of tactile stimulation, such as textured surfaces, edges, and vibration. And, of course, they respond to touches on the glabrous skin of the palm and fingers. Most important, and in noted contrast to the C-tactile fibers, A-betas are most effectively activated by intense stimuli: The more rapid the stroke, the stronger the response. A-beta fibers are capable of resolving a wide range of touch stimulus properties, while the C-tactile system appears to be tuned to detect a particular type of touch: a light caress within a particular range of speed. This tuning for an optimal speed is crucial for perception. When caresses of various speeds are delivered to the forearm or thigh of normal subjects, the caresses that are reported as feeling the most pleasant are those in the 3- to 10-centimeters-per-second range, which are precisely those that most strongly activate the C-tactile fibers.[7]

When brain-imaging studies are performed with normal subjects, a forearm caress evokes activation of the primary and secondary somatosensory cortices, which are involved in fine shape and texture discrimination and driven by information originating in A-beta fibers, as well as a region called the posterior insula, which has been implicated in the emotional aspects of sensory processing. When G.L.'s brain is imaged, the forearm caress activates the posterior insula, but not the primary or secondary somatosensory areas, suggesting that her spared C-tactile fibers strongly activate the former but not the latter (figure 3.3). Furthermore, the same intermediate-speed caresses that strongly activate C-tactile fibers and are rated as the most pleasant also produce the strongest activation of the posterior insula in both G.L. and uninjured subjects.[8]

Taken together, these experiments argue that C-tactile fibers function as caress detectors that innervate the hairy skin and project through a pathway to the posterior insula, the activation of

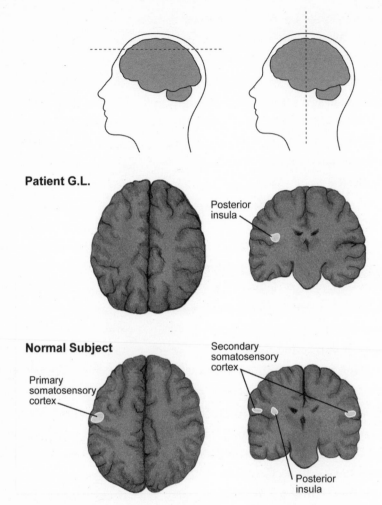

Figure 3.3 A light caress on the right forearm activates the left anterior insula in both patient G.L., who lacks large myelinated sensory nerve fibers, and normal, uninjured subjects. The primary and secondary somatosensory cortices, which are necessary for fine discrimination, recognition, and localization of touch stimuli, are not activated by a forearm caress in G.L. Note that, consistent with our discussion in chapter 2, stimulation of the right forearm activates the left primary somatosensory cortex, but both left and right secondary somatosensory cortices in the normal subjects. The data depicted here are from H. Olausson, Y. Lamarre, H. Backlund, C. Morin, B. G. Wallin, G. Starck, S. Ekholm, I. Strigo, K. Worsley, Å. B. Vallbo, and M. C. Bushnell, "Unmyelinated tactile afferents signal touch and project to insular cortex," *Nature Neuroscience* 5 (2002): 900–904, with permission of Nature Publishing Group.

which produces a slow, diffuse, pleasant sensation. Importantly, this pathway functions in all people, not just those suffering from sensory neuronopathy, like G.L. That includes the nineteen-year-old rapist in my jury duty case. While the glans penis is glabrous, the shaft of the penis is hairy skin and thereby innervated by C-tactile afferents. It's likely that the handjob that enraged him by being either too fast or too slow was outside the range that would strongly activate the C-tactile sensors in the hairy skin of his penile shaft.

~~~~~

If you ask people in Norrbotten, a large, sparsely populated region of Sweden located north of the Arctic Circle, they'll tell you that the problem started in the seventeenth century or perhaps even earlier, with a man who couldn't feel pain. Consequently, he sustained frequent injuries, ranging from skin abrasions to broken bones to deteriorated joints, a trait he passed on in his family line.[9] Over the years it has not been uncommon for the residents of Norrbotten to marry their first or second cousins, reinforcing hereditary pain insensitivity, which has continued in the region to this day.

The Norrbotten pain-insensitive patients are affected with the condition to varying degrees, but all tend to have severely reduced sensations of both deep and surface pain as well as temperature. They are cognitively normal and their sense of fine touch is unimpaired, as is their proprioception and motor coordination. Genetic testing has revealed that the Norrbotten mutation disrupts a gene that encodes a protein called nerve growth factor beta (NGFbeta). Because NGFbeta is required for the survival of small sensory neurons, it is not surprising that nerve biopsies of these patients reveal a loss of C-fibers and A-delta fibers, sparing the larger myelinated A-alpha and A-beta fibers (figure 3.2).[10]

The Norrbotten sensory nerve syndrome is almost exactly the opposite of that displayed by patient G.L., whose C- and A-delta fibers survive. While the Norrbotten patients are best known for their pain insensitivity, their loss of C-fibers means that their caress-detecting C-tactile fibers are impaired as well. When Norrbotten patients are evaluated in a brain scanner, a forearm caress at the ideal intermediate speed produces only a weak activation of the posterior insula, and they rate the caress as significantly less pleasant than do age- and sex-matched control subjects.[11] Putting all the puzzle pieces together—the results from C-tactile fiber-lacking Norrbotten patients and C-fiber-spared patients like G.L.—suggests that C-tactile fibers drive a specialized system in the brain for sensing caresses.[12]

~~~~~

Thus we have two separate touch systems in the skin, operating in parallel, which report fundamentally different aspects of our tactile world (figure 3.4). The fast A-beta fibers convey touch information with high spatial and temporal resolution, which allow one to discriminate between subtly different stimuli anywhere on the body. They are all about the facts. The slow C-fiber caress system produces a diffuse, emotionally positive sensation on only the hairy skin. It is all about the emotional vibe, and as such conveys both the crucial social information that's necessary for the proper emotional development of newborns and the social touches later in life that are critical for the development of trust and cooperation, in both humans and other animals.[13]

Why is it necessary to have a slow, diffuse C-tactile fiber system at all? Isn't the information conveyed by the C-tactile fibers merely a subset of that already encoded in the A-beta fibers? Why not just detect caresses with A-beta fibers and fast mechanosensors? One possible answer: If we imagine that the emotional information of social touch is determined by stroke velocity, then it may be easier to have dedicated slow-velocity detectors like the C-tactile fibers. In wide-range detectors, like those innervated by A-beta fibers, the information encoding emotional significance is buried in other tactile signals that do not have emotional meaning and is therefore difficult to extract. Well, then, why not just have a subset of fast

A-beta fibers that are tuned for optimal caressing speeds? That would give you the best of both worlds—you could detect caresses selectively but maintain fast transmission of information. There is no definitive answer to this question. It may be that the temporal integration produced by having slow signals is actually better for making decisions based upon emotional touch. These might be the kinds of choices that you want to make more deliberately, based on a longer touch stimulus. You don't want to misinterpret a casual brush, for example, as a socially motivated caress. (As my teenage kids would say, "That would be so awkward.") Or it may be that A-beta fibers are simply too expensive, in the respect that they use a great deal of energy, and their myelin wrapping needs a lot of cross-sectional space in the nerve. If you don't need the speed, it's advantageous not to pay the biological price for it and instead to use a cheap, slow fiber. Alternatively, it may simply be that the C-fibers evolved first, and that this early system constrained further evolutionary change.

It's important to note that these two touch systems, the fast and the slow, are not completely separate. There is two-way communication between the posterior insula, the main cortical hub of the C-tactile caress system, and the primary and higher somatosensory cortices, the cortical regions that process fine tactile discrimination from A-beta fibers (figure 3.4). Each may influence the other, and the entire system is under powerful multisensory and emotional modulation relating to situational and social contexts. The exact same optimal-velocity caress may feel entirely different coming from a sweetheart than it does if applied by a stranger—or even from a sweetheart during a loving, connected time versus in the middle of an unresolved argument. While the strongest activation by an optimal caress is seen in the posterior insula, recent work has shown that the primary somatosensory cortex can also be activated by caress and that the degree of activation can be modulated by social-cognitive factors like the perceived sex of the caresser, information that is presumably received from nonsomatosensory brain regions.[14]

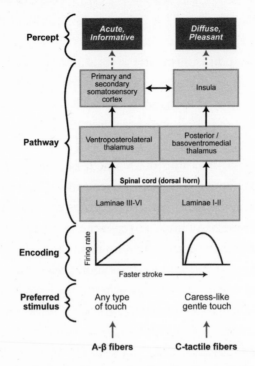

Figure 3.4 A schematic diagram showing the flow of information in two separate pathways from the hairy skin to the brain that mediate fast informative touch and slow diffuse caress detection. The A-beta system responds to any type of touch in a linear fashion: Faster and more forceful strokes result in stronger electrical signals (more spike firing) in the fast myelinated axons. These fibers ascend in the deeper layers of the spinal dorsal horn, called laminae III–VI, and, following a synaptic relay in the brain stem, cross the midline and activate a region of the thalamus called the ventroposterolateral nucleus. These thalamic neurons then send their axons to activate the primary and higher somatosensory cortices, where successive computations underlying discriminative fine touch are performed. By contrast, the C-tactile system is tuned for light stroking at intermediate speed, imposing a strong initial filter on tactile information. The slow C-tactile fibers contact neurons in the more superficial lamina I and II of the spinal dorsal horn, and these cross the midline and send their axons along a pathway called the spinothalamic tract to activate neurons in different regions of the thalamus: the posterior and basoventromedial nuclei, which in turn send their axons to the posterior insula, where the pleasant diffuse percept is felt. This figure is adapted from I. Morrison, L. S. Löken, and H. Olausson, "The skin as a social organ," *Experimental Brain Research* 204 (2010): 305–14, with permission of Springer.

In chapter 1 we discussed the important role of social touch in developing and reinforcing trust and cooperation in a wide range of bonding situations, ranging from babies to adults and from colleagues to lovers. A caress communicates that you are safe. You can trust the person administering it, just as you trust your mother, who first caressed you; he or she is not a threat. The C-tactile system plays a crucial role in this communication. Caresses activate not only the posterior insula and the somatosensory cortices but also other brain regions that integrate many kinds of sensory and motor information. These include areas of the cortex involved in social cognition, such as the superior temporal sulcus, the medial prefrontal cortex, and the anterior cingulate cortex.[15] Of course, studies of such regions were conducted with normal subjects, not with patients like G.L., so these social cognition centers likely received some fast A-beta tactile signals as well. However, they showed significantly less activation in response to a fast caress than to an optimal-speed caress, consistent with an important role for the C-tactile system in driving their responses, and, presumably, social bonding.

Adults with autism spectrum disorder have a range of challenges in social cognition and struggle to discern the social intentions of others. They tend to have an aversion to certain forms of social touch and rate optimal caresses as less pleasant than matched control subjects do. Furthermore, there is a positive correlation between the severity of autism and the reduction in perceived pleasantness: Patients with the most severe autism gave the lowest caress ratings of all. When brain imaging was performed on autistic subjects, a similar correlation emerged: Those with the most severe autism had the smallest activation of certain social cognition centers (the medial prefrontal cortex and the superior temporal sulcus) by an optimal-speed arm caress.[16]

This study, while provocative and interesting, leaves much unexplained. Where is the defect in caress processing actually located?

Do the C-tactile fibers of severely autistic people encode caresses normally, or does the defect exist in the activity of the skin and sensory nerves? Is the response to caresses in the posterior insula normal in autistic people? And, perhaps most important, what is the causality here? Is there a defect in caress sensation that makes autistic people averse to social touch? And does this make it harder for them to discern the social intentions of others? Recall that the Norrbotten patients, who lack C-tactile fibers, appear to be cognitively normal and do not show signs of autism (although this particular question has not been studied carefully). Alternatively, perhaps caress seeking and caress liking require a certain amount of experience to develop. Autistic people, by living a less social life for reasons unrelated to touch, might fail to have those experiences early in life.

When I was thirteen years old, like most people, I had yet to have any significant romantic experience. So I was very attentive when my mother took me to see the sexy (and twisted and political) film *Swept Away*, directed by Lina Wertmüller. Watching the leading couple, played by Giancarlo Giannini and Mariangela Melato, caress each other's skin was so intense, compelling, and immediate that it was almost more than I could stand. I wasn't just watching the actors—I felt their sensations on my own skin and I can clearly recall them today, nearly forty years later.

That we can thrill to the touch of performers on the screen as we might react to a caress on our own skin results from the neural information received by the posterior insula, the main cortical center activated by caresses and a crucial node in the emotional brain. In addition to C-tactile fibers, the posterior insula receives highly processed visual information as well. Amazingly, merely watching a movie of someone's arm being caressed will activate the posterior insula of a subject in a manner that's similar to her receiving a real arm caress. Even more surprisingly, just as with a real tactile caress, the posterior insula is most strongly activated by a movie showing optimal caressing speed as opposed to faster or slower stimulation, and the viewer herself rates optimal-speed caress movies as most pleasant. Norrbotten patients, who lack C-tactile fibers, rated the caress footage as significantly less pleasant than control subjects and had no particularly sensitivity to caress speed.[17] In this way, both the normal subjects and the Norrbotten patients evaluated the caress film in a way that was anchored in their own tactile experiences of caresses.

We humans are tuned to crucial features of emotional touch not just in our own tactile experience but also when observing others. We are highly sensitive to reading such signals between other people. This is an important feature of social cognition, helping us to track changes in affiliations, coalitions, and status within our social groups. And, of course, it is the source of endless gossip: Did you see how she touched him on the arm?

CHAPTER FOUR

SEXUAL TOUCH

It was early on in our relationship. B. and I had slept together only a few times and were very much in the discovery phase about each other's bodies and sexual likes and dislikes. After a wonderful night of lovemaking, we had drifted off to sleep and stirred only many hours later when the bright, annoyingly persistent sunshine slanted through the gaps in the window curtain. We began nuzzling and mumbling, half asleep, happy, and hazy. The bed smelled reassuringly funky from the previous night's activities, and the olfactory cloud added to the befuddlement of waking to render the scene unusually dreamlike. As we kissed I slowly moved my hand over her belly and upward to softly cup her breast. She made an encouraging purr, so I began to gently roll her nipple between my thumb and forefinger. This didn't seem to evoke any reaction at all, which struck me as a bit unusual, as she had previously responded to that kind of touching—sometimes even achieving orgasm from that alone. Her nipple, which also felt odd between my fingers, suddenly detached entirely from her breast, and I was left holding it in my hand.

At that precise moment the world ceased to make sense, so I tried to take stock of the situation:

a. I was now holding B.'s disembodied nipple in my hand.

b. The fact that I was doing so didn't seem to faze her at all. She was giving me a sleepy smile, though it quickly changed into a look of concern in response to my horrified expression.

c. There was no blood.

My thoughts were in overdrive; I was gaining no cognitive traction at all. How could this possibly be happening?

In a conventional tragic incident, like a devastating car wreck, our lives are changed in an instant, but the event conforms to a brutal, predictable physics. Objects collide and dissipate force. Gravity and friction exist. However much we are shaken, our core assumptions about the physical world are not violated. But lying there in bed, late on a sunny morning, I was at once the illusionist and the audience, and I had completely stumped myself. Given my previous life experience up to that moment, all three of the above observations could simply not be true.

Even today I have no sense of the actual duration of that event, and it's likely that only a few seconds elapsed before I continued to stroke the detached nipple. Then I began to notice that it wasn't the familiar warm, soft, wrinkly bud but was larger than usual, spongy and faintly waxy, and, of course, upon continued exploration I realized it was not a nipple at all. The spinning wheels of my mind finally found purchase: The spongy object was in fact a foam earplug that had come loose in the night and landed on her chest. My heart was racing, and I sputtered with relief, though it would be many more seconds—whole cognitive lifetimes—before I could explain to B. what had happened.

~~~~~~

The detached nipple story illustrates an important point about sensation, perception, and the difference between the two. The way we perceive a sensory event is not determined simply by the physical parameters of the stimulus involved (for example, 10 grams of force delivered to the pad of the thumb and index finger). Nor is it accounted for by how those stimulus parameters are filtered by the receptors that transduce it (in this case, the response properties of the skin mechanosensors in the pads of that thumb and index finger). Even when we add to this data additional information provided by our exploratory behavior (proprioceptive signals from the muscles in the hand and arm), it's *still* not enough to account for our ultimate perception of the stimulus. Rather, our perception of a sensory stimulus is crucially dependent upon our expectations, as they have been formed by our life experience up to that moment.[1] We know that nipples don't just fall off, and if somehow they could detach, we would expect that bleeding (and howls of pain) should result. We are confident from what we've learned that gravity should operate, living mammalian bodies should be warm, and so on. When there's a mismatch between expectation and sensation, it's a sign that something weird is happening, and our perception of that sensation is fundamentally altered.

Likewise, context is key in sensory experience. For most of us, the feeling of a finger tracing our lips is delightful and arousing in a romantic setting with a lover but decidedly unerotic when it takes place in the doctor's office.[2] And, of course, contextual perception is not limited to the sense of touch or to sexual situations. The taste of coffee can be jarring if you were expecting tea in your mug, even if you typically enjoy coffee. A faintly sulfurous odor lingering in a public toilet is disgusting, but that same odor can be delightful when encountered in a cheese shop (figure 4.1).

There are all kinds of sexual touching—lips and nipples often play key roles. For many, light touching or kissing of the ears, neck, inner arm, or anus can be sexually transporting. It doesn't matter if you're male or female or intersex, gay or straight, or bisexual—if you ask enough people in any group, you will find someone who experiences a sexual sensation from being touched on almost any part of the body. (If you're not convinced of this, just search the Internet for

Figure 4.1 The same odor can be perceived as foul or pleasant, depending upon context and expectation. This is a general feature of perception that also applies to tactile sensations. Comic by Julia Wertz; used with permission.

"eyebrow fetish" or "armpit sex.") The near-limitless variation in sexual behavior is a central feature of our human experience.

Still, while giving armpits, ears, and eyebrows their due in erotic potential, the genitals are unique. In the right context almost everyone finds stimulation of the external portion of the clitoris or the glans of the penis to be sexually exciting. Why is that the case? Is there something special about the structure of the skin in those locations? Like the lips and the fingertips, the skin of the external clitoris and the glans penis is glabrous, lacking even the finest hairs and their associated nerve fibers. As we've seen, the lips and fingertips are richly endowed with mechanoreceptors that allow for fine discriminative touch, but both the clitoris and glans penis have very few of these. What these tissues do have in abundance are free nerve endings that transduce heat, cold, pain, and inflammation.[3] In addition, they have a specialized type of nerve ending consisting of a coiled axon wrapped by a few nonneuronal encapsulating cells (figure 4.2). These have been called genital end bulbs (or, in the wonderful original German term, Genitalnervenkörperchen).[4]

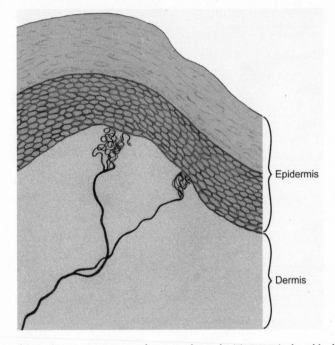

Figure 4.2 Is this structure important for sexual touch? Two genital end bulbs
in the skin of the human glans penis, located in the most superficial layer of
the dermis. These genital end bulbs are also found in the glans clitoris at even
higher density.

While there's been some argument about their distribution over
the years, it seems that the genital end bulbs are not entirely restricted
to the genitals. In other tissues that are similarly coiled, loosely
wrapped nerve endings are called mucocutaneous end-organs. They
are also present in other types of glabrous skin that can be involved
in sexual touch, such as the lips, the nipples, and the skin surrounding the anal opening. Perhaps the external clitoris and the glans
penis play a prominent role in sexual sensation merely because they
have a high density of this type of nerve ending. The external portion of the clitoris has the highest density of any tissue, and within
the glans penis, the greatest density of genital end bulbs is found at
the corona (the ridge marking the edge of the glans) and the frenu-
lum (the elastic tissue on the underside of the glans penis). These

are spots that men often report to be the very most sensitive locations for sexual stimulation.[5]

But do mucocutaneous end-organs truly have a special role in sexual touch sensation? Unfortunately, we don't know, as there are as of yet no drugs or genetic tricks that can selectively activate or inactivate them, in either humans or laboratory animals. We don't even know if their electrical signals follow a unique path into the spinal cord or the brain. In fact, to my knowledge, their electrical signals have not even been recorded in isolation. For a structure that's potentially so important to human life, it's amazing how little we understand about it.[6]

~~~~~~

Touch sensations from the genital region pass through three different nerves on their way to the spinal cord and, ultimately, the brain (figure 4.3). The pudendal nerve is the most important for sexual sensation, carrying signals from the clitoris in women and the penis in men. Because the penis and the clitoris derive from the same undifferentiated bit of embryonic tissue (which ultimately takes its particular form in males and females based upon the influence of sex hormones during early development), it is not surprising that they are innervated by the same sensory nerve. Importantly, a single sensory nerve can carry information from several different parts of the crotch. In women, the pelvic nerve conveys touch signals from the labia minora, the vaginal walls, the anus, and the rectum. In men, the pudendal nerve carries information from the anus and the scrotum as well as the penis. In women, sensations from the cervix and the uterus can also be conveyed by the hypogastric nerve as well as the cranial nerve called the vagus nerve, which travels directly to the brain stem, thereby bypassing the spinal cord entirely.

This convergence and divergence of touch signals from the pelvic region have some important consequences for our experience of sex. Cross talk between adjacent genital and perigenital locations may help explain why stimulation of the anus, rectum, and perineum can sometimes elicit strong sexual sensations. While many people enjoy stimulation of these areas during sex, some find that perigenital erotic sensations are not entirely limited to sexual situations. For example, a man in the Netherlands reported that he had an orgasm every time he would defecate.[7]

The wiring diagram of the sensory nerves also helps us understand how sexual sensation can be preserved in some cases of spinal cord injury. An injury resulting in complete transection of the spinal cord in the lower back, at the level of the second lumbar vertebra, will interrupt sensations from the penis and scrotum in men and the clitoris and labia minora in women, as carried by the pudendal and pelvic nerves. However, because the hypogastric and vagal nerve signals still have an intact path to the brain, this damage to the spinal cord would spare some sexual sensations from the cervix and uterus in women and the testicles and prostate in men. Indeed, some women with complete spinal cord transection even higher in the spinal column, well above the entrance of the hypogastric nerve, report strong sexual sensations from the cervix and uterus, which are thought to be carried by the vagus nerve.[8]

It's easy to look at schematic anatomical drawings like that in figure 4.3 and imagine that the wiring diagram for the nerves carrying sensations from the pelvic region is absolutely the same in every man and every woman. At the broadest level, this is true: Nearly every man will have sensation from the penis, and nearly every woman will have sensation from the clitoris, carried by the pudendal nerve. But when we look at the fine branching patterns of the nerves and the distribution of the nerve endings, individual variation becomes apparent. Comparing two women, one might have somewhat fewer genital end bulbs in her external clitoris but more in her labia minora and anus. Comparing two men, one might have more free nerve endings in his prostate but fewer in his scrotum.

Could the fine structure of the sensory nerves innervating the genital and perigenital regions underlie individual variation in preferred sexual activities? This idea has been suggested by anatomists

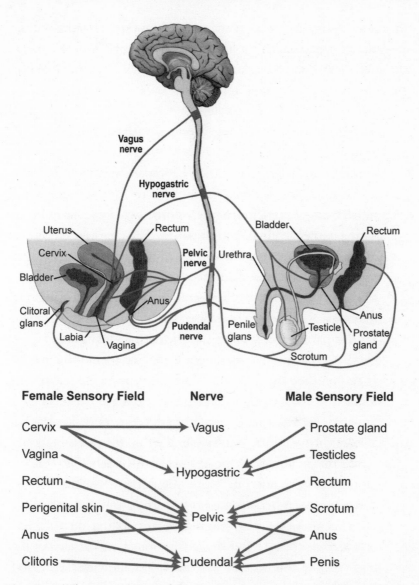

Figure 4.3 The organization of the sensory nerves innervating the pelvic region. Sensory information from the pelvis is carried to the brain by three pairs of spinal nerves—the pudendal, the pelvic, and the hypogastric—which enter at different levels along the spinal cord. Sensations from the uterus and cervix are also conveyed via a cranial nerve, called the vagus, which enters the brain stem directly. It is important to note that even a single nerve can convey information from a number of different skin sources. The male pudendal nerve, for example, carries sensations from the penis, anus, and scrotum. © 2013 Joan M. K. Tycko

for many years.[5] More recently, feminist author Naomi Wolf has extended this argument. In her book *Vagina: A Cultural History*, she speculates that individual neural variation can explain differences in women's sexual preferences:

> For some women, a lot of neural pathways originate in the clitoris, and these women's vaginas will be less "innervated"—less dense with nerves. A woman in this group may like clitoral stimulation a lot, and not get as much from penetration. Some women have lots of innervation in their vaginas, and climax easily from penetration alone. Another woman may have a lot of neural pathway terminations in the perineal or anal area; she may like anal sex and even be able to have an orgasm from it, while it may leave a differently wired woman completely cold, or even in pain. Some women's pelvic neural wiring will be closer to the surface, making it easier for them to reach orgasm; other women's neural wiring may be more submerged in their bodies, driving them and their partners to need to be more patient and inventive, as they must seek a more elusive climax. Culture and upbringing definitely have a role in how you climax and can affect whether you climax easily or not, but that is not all there is to it. This discourse heaps vast unnecessary guilt and shame on millions of women or, conversely, depending on their tastes, leads them to feel slightly perverted. . . . Whatever it is you like and need in bed—as a woman, with all that variability—these preferences may just be due to your physical wiring.[9]

This is a reasonable proposition: Sensory nerves in this region of the body do transmit sexual sensation, and there is individual variation in the fine structure of these nerves. So it is plausible that subtle differences in the configuration of the nerves of the pelvis might underlie at least some portion of the individual variation in sexual experience and therefore preferences for certain types or styles of sexual act. However, several caveats must be raised. Most important, there is no evidence to *prove* that normal variation in the fine structure of genital sensory nerves underlies differences in sexual sensation or sexual preferences. The anatomical variation involved here is below the resolution limit of today's medical scanning machines, and the knowledge we do have of the fine structure of the relevant sensory nerves and their endings in the skin and other tissues comes from cadavers and biopsy samples, which we can cut into thin slices and examine under a microscope, not from healthy, intact living people whom we can interview about their sexual experiences.[10]

Sexual sensation and sexual desire require an ongoing dialogue between the body (particularly the genitals) and the brain. When considering potential biological substrates for individual differences in sexual experience, we should also include variation in those parts of the brain activated by sexual touch. Crucially, in both the skin and the brain, that variation might not be only structural in nature. The most significant individual differences might be a result of the electrical or chemical signaling functions of neurons—features that do not necessarily involve changes in the shape of neurons or their wiring diagram. Differences in the properties of ion channels or neurotransmitter receptors can have a profound effect on the function of a neuron in a sexual touch circuit, but these differences cannot be identified by structural measurement, even with the most powerful microscope. So when we read Wolf's concluding phrase, "these [sexual] preferences may just be due to your physical wiring," we should remember that the word "may" is key, as the causal link remains unproven. And we should also think of "wiring" in the broadest sense of the term: not just the wiring of nerves in the genital and perigenital regions but also the wiring of neurons in the relevant parts of the brain. Additionally, we should also include individual variation that changes the electrical and chemical signals of neurons but does not change the neuronal wiring diagram (in either the brain or the skin) at all.[11]

~~~~~~

What can be learned by mapping the pattern of brain activation in the somatosensory cortex resulting from genital stimulation? Touch signals from the pelvis are carried to the brain along different pathways in the spine and the brain stem, depending upon the touch modality (fine discriminative touch, caress, temperature, etc.). Like other regions of the body, pelvic tactile signals form a relay connection in the thalamus and then arrive at the neocortex, where they are represented in a body map in the primary somatosensory region (figure 2.8). As we discussed previously, the genitals occupy a fairly small section of the overall body map, consistent with their low density of mechanosensors. In one study, undertaken by Barry Komisaruk and his colleagues at Rutgers University, women in a brain scanner were given a handheld dildo and asked to self-stimulate various genital regions—the external clitoris, the vagina, and the cervix.[12] When examining the pattern of activation in the primary somatosensory cortex, each of these three genital regions activated a discrete patch of brain tissue, but these patches were adjacent to one another, and in some cases partially overlapped (figure 4.4). Importantly, and consistent with Penfield's maps derived from brain stimulation, there is some individual variation in the size and precise location of these patches. Might the size of the patches be determined by individual variation in the fine structure of sensory nerve endings in the pelvis? If a woman has an unusually dense sensory innervation of the vaginal walls or a man has a particularly dense innervation of the scrotum, will those regions be enlarged in the body map? And what about the effects of experience? If one regularly engages in anal sex, will that expand the cortical representation of the anus and rectum in much the same way that daily violin practice expands the sensory map of the fingering hand? These questions remain to be addressed.

Consistent with Penfield's classical body map (figure 2.8), the stimulation of the female genitals activated an area detached from the contiguous portion of the map, beyond the representation of toes. However, examining figure 4.4 you can also see adjacent, semi-overlapping patches of activation just where you might expect, at the groin portion of the map at the intersection of the thighs and the abdomen. Why are the female genitals represented in two places on the map? It turns out that this is not only a female phenomenon, as men have dual representation of the genitals as well. Figure 4.4 shows activity in both locations, but the authors of this study claim that the patches near the toes are the real genital activation sites and the patches at the groin site result from incidental activation of the adjacent tissue surrounding the genitals.[13]

Interestingly, when women in this experiment stimulated their nipples, two different patches of activation were likewise seen in the brain scanner: one at the chest site on the body map, and another detached one, located beyond the toes. In fact, the patch activated by nipple stimulation significantly overlapped those of the clitoris, vagina, and cervix. This may explain why nipple stimulation is sexually exciting to many women. More generally, it raises the issue of whether this particular area of the primary somatosensory cortex (called the mesial surface of the postcentral gyrus) has a special role in sexual touch.[14] If you were to examine a microscope slide prepared from postmortem tissue of this area of the brain, you'd find nothing unusual about it. The individual neurons and glial cells and their overall layered structure look almost the same as those in regions processing somatosensory information from less erotic parts of the body. It may turn out that this region has unusually strong connections with other brain regions involved in pleasure and fear, but to date this has not been investigated.

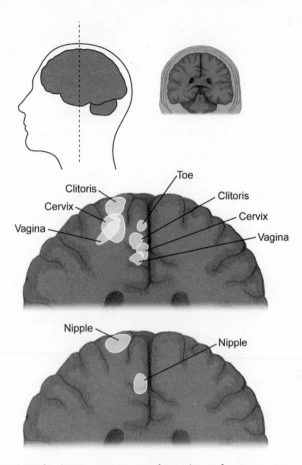

Figure 4.4 Using a brain scanner to map the regions of somatosensory cortices activated by self-stimulation of the external clitoris, vagina, cervix, toe, and nipple. Note that nipple stimulation produced activation that overlapped with that of the cervix, vagina, and clitoris. This might underlie the role of the nipple in women's sexual sensation. To date, this experiment has not been performed with male subjects. Another issue to bear in mind is that it's not possible to stimulate all genital regions in isolation using a dildo. Stimulation of the vagina will certainly activate the base of the clitoris, which presses against the anterior vaginal wall; stimulation of the cervix will undoubtedly involve some activation of both the vagina and the clitoris; and all of these forms of genital stimulation will produce some stretching of the labia minora and the perineum. Adapted from B. R. Komisaruk, N. Wise, E. Frangis, W.-C. Liu, K. Allen, and S. Brody, "Women's clitoris, vagina, and cervix mapped on the sensory cortex: fMRI evidence," *Journal of Sexual Medicine* 8 (2011): 2822–30, with permission of the publisher, Wiley.

~~~~~~

When I was in the fourth grade, the boy sitting at the desk next to mine was named Ralph. Even at ten years old, he was one of those kids who you knew was going to eventually wind up in prison. Ralph fought with everyone. He loved to draw sloppy pictures of motorcycles on his filthy arm with ballpoint pen and had a perpetually runny nose that he never wiped. One day, apropos of nothing, he leaned over and said to me, "You know how girls get pregnant? They take off their pants—*sniff*—and a boy takes off his pants—*sniff*—and then they rub their butts together. Then she'll have a baby." At that point in time the precise details of sex were still a bit vague to me, but I knew that this couldn't be right. Ralph continued with his sexual riff: "You know what a boner is? If you think about naked ladies—*sniff*—then your dick will get stiff, and that's called a boner." To me, this explanation seemed equally suspect. A boner, as I knew, was a tangible, concrete thing, while a thought—of naked ladies or anything else—was wispy and fleeting and clearly not of the physical world. How could one have anything to do with the other? They were in different realms. His explanation smacked of mysticism and, worse, superstition.

"No way," I said.

"Way," said Ralph.

~~~~~~

Let's imagine that you're walking down Waverly Place in New York City's West Village. You're hungry and a bit tired. You see someone walking by carrying a falafel sandwich, one of your favorite foods. Then, moving along, you spot the source of the falafel, Taïm, a tiny, crowded restaurant. You can smell the delicious odors of frying falafel balls, harissa, and tahini sauce, and you begin to salivate. You walk in and order a falafel sandwich. A few minutes later, you take that first crispy bite, which is so very good. You take another, and because it's a big sandwich, as you continue to eat, you begin to feel sated. The pleasure from the last bite is fine but not nearly as good as that first bite. Now you're full up and unlikely to eat again for a while. Unless, perhaps, there's some good sorbet to be had over on Prince Street.

In true nerdly tradition, if we were to graph this pleasurable eating experience (pleasure on the y-axis, time on the x-axis), it might look something like figure 4.5. There's an initial phase in which the sight and smell of food interact with your internal state (hunger) to

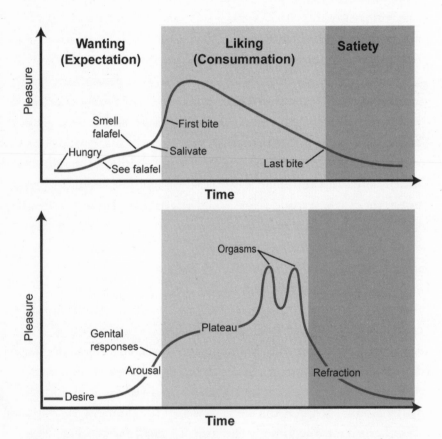

Figure 4.5 Pleasure graphs for eating and sex reveal some common features. This figure is adapted and extended from J. R. Georgiadis, M. L. Kringelbach, and J. G. Pfaus, "Sex for fun: a synthesis of human and animal neurobiology," *Nature Reviews Urology* 9 (2012): 486–98, with permission of Nature Publishing Group.

create desire. You're already feeling some pleasure at this point, and in anticipation of more you begin to salivate. Importantly, this desire reflects both hardwired and learned responses. We humans are hardwired to respond favorably to certain food odors, including those present in fried foods like falafel. In addition, our responses are conditioned by experience. You know that you like falafel; you've been to this restaurant before and have enjoyed it. (Conversely, if you got sick last week after eating a falafel, it's almost certain that its odors and the sight of it would not induce desire, even if you were hungry.)

Next is the liking phase, when pleasure peaks. That first bite gives a surge of pleasure, and each subsequent bite, somewhat less. In part this is because you are habituating to the smell, taste, and mouthfeel sensations of the food. The first bite is rather novel; subsequent bites become increasingly familiar. Somewhat later you begin responding to a mixture of conscious and subconscious signals from your body telling you that your stomach is distended and the nutrient levels in your blood have increased. This leads to the satiety phase, in which you're less likely to eat more falafel, but you might be persuaded by something new, like a cup of fruity sorbet.

I've gone off on this food tangent to make a point—namely, that sexual activity is not a singular form of human experience. While it has certain unique aspects, in many ways it's not that different from other pleasures, like eating. Let's imagine a similar narrative concerning sex. Here I've chosen an example involving two women, but the particular actors involved don't matter that much. Let's imagine that you're about to ovulate, and are thus at the time of the month when sexual desire runs highest. You're feeling energetic and playful, lying on the sofa reading the newspaper. Your sweetheart comes through the door. She looks great, vibrant and slightly flushed from exercise. She comes over for a kiss, and you can smell her body and her hair and feel her soft lips brush your own. You begin to feel aroused as you realize that she is receptive. Your vagina starts to lubricate, your muscle tone softens, and you continue to kiss and caress.

Clothes come off, and you can see her body. Then things really begin: kissing, stroking, licking in lots of places, with particular attention to the genital and perigenital regions. There's a slow buildup of tension, an inevitable feeling of progression, and then orgasm, sometimes one, sometimes many. When you're done, lingering in the afterglow, you're less likely to initiate sex again immediately, unless something unusual happens—like your sweetie suggests some sexual activity that you've always wanted to try but have never done.

The pleasure graph for this experience is shown in figure 4.5 below the falafel graph, and it's evident that there are many similarities. In both cases an internal state (hunger, a sexually receptive time in the ovarian cycle) primes you to respond to appealing sights, smells, and sounds. In both cases your mounting desire in the wanting phase is mediated by both hardwired responses and your own past experience. (You may be remembering a previous pleasurable sexual encounter.) And that increasing desire is accompanied by changes in your body that you can't control (salivation, vaginal lubrication).

In the liking phase we begin to see some divergence in the pleasure graphs. Typically, the first bite of food gives you the most pleasure, while the first sexual touch does not. Most often there's a buildup to orgasm, and it's orgasm itself that is the most pleasurable moment of sex. And there's more individual variation in sexual liking: You might reach orgasm quickly, while others achieve it slowly or not at all.[15] You might typically have one orgasm in a sexual encounter, while others have many. As we have discussed, the variation in preferred sexual behaviors is huge and, of course, is not limited to the details of touching. And, of course, both eating and sex tend to produce satiety, which can be overcome by particular circumstance, most notably novelty.[16]

~~~~~~

So that kid Ralph was right: Merely having sexual thoughts can prepare the genitals for sexual activity. But that's not the whole story. Penile erection and vaginal lubrication can result either from signals descending from the brain during sexual thoughts or from genital or perigenital touching. Most often, in either partnered or solo sex, sexual thoughts and genital touching go together, and both contribute to genital preparatory responses. But this is not always the case. If spinal cord injury interrupts the neural pathways from the brain to the genitals, then sexual thoughts will be unable to evoke erection or lubrication. However, if the sensory pathways from the genital or perigenital regions to the spinal cord are intact, reflex circuits in the spinal cord are sufficient for genital touching to evoke lubrication/erection. This is true even if those touches cannot be perceived due to damaged spinal cord fibers carrying touch information up to the brain.[17]

Genital responses are one area where men and women differ considerably. For the most part, men get erections only when they are sexually aroused or when their genitals are directly stimulated (sometimes in a nonsexual way—by clothing, for example). However, when women are fitted with a tampon-sized probe that measures vaginal lubrication, it is revealed that they often lubricate in response to sensory stimuli that they report not to be sexually arousing. In one study, most of the straight women lubricated in response to videos of woman-woman or man-man sex (or even bonobo-bonobo sex), although they reported that they were not consciously aroused by these images. Likewise, most lesbians lubricated in response to male-female or male-male sex videos, even when they, too, reported not being aroused by them.[18] Of course, I don't mean to imply that there are no straight women who are aroused by male-male sex or female-female sex or even bonobo sex, and there are certainly plenty of lesbians who find male-female and male-male sex videos sexually exciting. The point is that, on average, males—gay, straight, or bi—tend to get erections only in response to stimuli (or thoughts) that they report as arousing while, on average, women—gay, straight, or bi—get wet in response to a much wider range of sexual stimuli, including those that they specifically report as not being arousing. Sex researchers Meredith

Chivers and Ellen Laan have both proposed that reflexive vaginal lubrication in response to a broad array of sexual stimuli is an adaptive response to situations in which penis-vagina sex is rapid or nonconsensual: Lubrication would reduce the chance of vaginal injury or infection. They speculate that this might have been the case for much of human evolutionary history.

Both men and women can suffer from genital responses in the absence of sexual desire. In men, this condition, in which an erection lasts from hours to many days and is not relieved by orgasm, is called priapism. It can occur as a side effect of many different maladies, including leukemia, sickle-cell disease, and pelvic tumors. Many drugs, both therapeutic (certain antidepressants and blood thinners as well as drugs used to treat erectile dysfunction) and recreational (cocaine and amphetamines), have also been linked to priapism. While the persistent erection of priapism is painful, it is not typically associated with a strong urge to stimulate the penis to achieve orgasm.

Like priapism, persistent genital arousal disorder (PGAD) in women results in vasocongestion in the genital area and consequent vaginal lubrication and swelling of the labia and clitoris. Unlike priapism, it is associated with hypersensitivity to touch. Innocuous stimuli like the movement of clothing or vibrations from riding in a car can cause a pelvic tingling sensation sometimes leading to orgasm. Most distressingly, PGAD often involves a strong, unwanted urge to masturbate (or otherwise achieve orgasm). It is not associated with increased sexual desire but rather is like a terrible itch that won't go away. An orgasm brings relief, but only briefly. PGAD sufferers do not enjoy their sexual sensations or compulsions and have reactions ranging from simple embarrassment (PGAD is almost certainly quite underreported) to deep distress about not being able to have normal social relationships, care for children, or hold down a job. In some extreme and tragic cases, women with PGAD have been effectively trapped in their homes, masturbating continually. Suicides are not unknown, like that of Gretchen

Molannen, who killed herself at her home in Spring Hill, Florida, at the age of thirty-nine after suffering unrelenting PGAD for sixteen years. In an interview with the *Tampa Bay Times* a few months before her death, she described her daily life:

> The arousal won't let up. It will not subside. It will not relent. One O-R-G will lead you directly into the horrible intense urge, like you're already next to having another one. So you just have to keep going. I mean, on my worst night I had 50 in a row. I can't even stop to get a drink of water. And you're in so much pain. You're soaking in sweat. Every inch of your body hurts. Your heart is pounding so hard . . . You have to ignore it, Gretchen. YOU DON'T HAVE A CHOICE. STOP NOW. Just let your body calm down. Many times, I've tried that. I'd be as far as in the bathroom, going in for my reward shower. I'm done. Now it's time to clean up and relax. And I'd look at myself in the mirror and there it is again. And I'd throw myself on the floor and cry. Men don't understand it. They don't care. They think it's hot . . . When I describe it to men, I tell them, "Imagine having an erection that does not go down, that feeling of just before it comes out, all day, all night, no matter how many times, no matter how much you've destroyed the skin on your penis."[19]

There is no single, well-defined cause of PGAD. In some cases it can be the result of the entrapment of the pudendal or pelvic nerves (or one of their branches) and can be relieved by surgery. In others it has been associated with vascular problems related to the flow of blood in the pelvis. A type of cyst that forms on the dorsal root ganglia in middle age, called a Tarlov cyst, is much more common in women with PGAD, suggesting a causal relationship. Some neuropsychiatric medications have been suggested as PGAD triggers, while others have been reported to alleviate it, but the scientific

literature on this topic is scattershot and confused. PGAD sometimes co-occurs with restless leg syndrome, which involves a similar
itchlike compulsion to move one's limbs, particularly the legs, and
which is also poorly understood. To date we don't know if PGAD
is associated with changes in the nerve endings in the skin of the
genital region, nor do we know what the pattern of brain activity
looks like in women who are feeling the PGAD-driven compulsion
to achieve orgasm.[20]

~~~~~

Sometimes the most familiar things resist description. In this vein,
let's consider the question, What exactly is an orgasm? Physiologically speaking, orgasm in both men and women involves an increase
in blood pressure, involuntary muscle contractions in the pelvis and
elsewhere, a rising heart rate, and an intense feeling of pleasure, followed by satisfaction. But that's a rather dry and clinical description, lacking in poetry. John Money, a Johns Hopkins psychiatrist
who was a pioneer of sex research, writing with colleagues, did a
good job of capturing both the transcendent and the biological in
this definition of orgasm:[21] "The zenith of sexuoerotic experience that
men and women characterize subjectively as a voluptuous rapture
or ecstasy. It occurs simultaneously in the brain/mind and the pelvic genitalia."[22] Money's description highlights several key points.
First, orgasm is a unique experience: It's not merely a more intense
form of touch sensation but a qualitatively different one. Second,
the most reliable and typical way of achieving orgasm is when touch
signals from genital stimulation are carried through sensory nerves
to the spinal cord and the brain. Third, orgasm ultimately occurs in
the brain, not the genitals.

   Can orgasms occur even with no involvement of the genitals?
Absolutely.[23] Some people can achieve orgasm from touching areas
of skin that are far from the pelvis, like the nipple, neck, mouth,
and even such ostensibly nonerotic areas as the nose and the knee.[24]

And of course, both men and women can have orgasms from stimulation of the anus and rectum, but this is likely due to cross talk between the genital and perigenital regions, as we've discussed (figure 4.3). In rare cases, even touching of any kind may be dispensable. Some people appear to be able to have orgasms through thought alone, or through ritualized breathing. And, of course, orgasms can occur during sleep, even when the genital area is not in contact with bedclothes. People with complete spinal cord injuries, who cannot feel their pelvic regions at all, report orgasms during dreaming sleep that feel as if they were occurring in the genitals.[25]

~~~~~

A group of men and women were asked to write paragraph-length descriptions of their own orgasms. Then the paragraphs were edited to remove any words (like penis or vagina) that would give a clue to the writer's sex. When the redacted paragraphs were given to a panel of judges (composed of medical students, psychologists, and gynecologists) to analyze, the descriptions of male and female orgasms were indistinguishable.[26] There are some important differences between them, however. On average, women's orgasms, as measured using a sensor that detects involuntary contractions of rectal muscles, are somewhat longer than men's (about twenty-five seconds, compared with fifteen seconds). Women are also much more likely to achieve orgasm more than once during a sexual encounter.

Sigmund Freud pronounced that while a young woman might achieve orgasm through clitoral stimulation, a mature woman had orgasms only when her vagina, but not her clitoris, was stimulated. He did not base this notion on anatomical or physiological measurements of the vagina or clitoris or the nerves that innervate them. Rather, he was motivated to construct a narrative in which penetration by a man's penis was crucial for a woman's sexual satisfaction. For their part, however, women are well aware that for

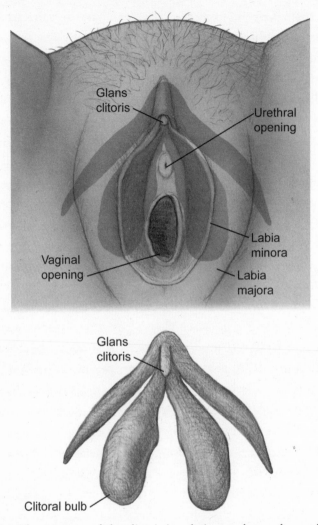

Figure 4.6 The anatomy of the clitoris in relation to the urethra and vagina. Top: The external portion of the clitoris (the clitoral glans) is a small part of the entire structure shown with shading. Importantly, the base of the clitoris contacts the anterior wall of the vagina. Bottom: Here, the entire clitoris is shown with other structures removed to clearly reveal its shape. In a chronic inflammatory skin disorder called lichen sclerosus, the clitoral hood can sometimes fuse with the surrounding labia minora to completely cover the external clitoris (similar fusion of the labia can partially or completely obstruct the vaginal opening). Even when the external clitoris is overgrown with skin, most women with lichen sclerosus are still able to achieve orgasm through indirect clitoral stimulation.[27] © 2013 Joan M. K. Tycko

most of them, young or mature, stimulation of the external clitoris, either during solo or partnered sex, is the most reliable path to achieving orgasm. This is not surprising: As we have discussed, the high density of free nerve endings and genital end bulbs in the external clitoris is consistent with its special role in sexual sensation.

Many years later the reaction against Freud's "vaginal orgasm" was vociferous, reaching its peak with the publication of "The Myth of the Vaginal Orgasm" by feminist scholar Anne Koedt in 1970.[28] Here the pendulum of thought swung far in the other direction. Koedt and others claimed that the clitoris was the only structure capable of transmitting women's sexual sensation and thereby triggering orgasm. Koedt cited pioneering sex researcher Alfred Kinsey, who stated that "[the vagina is] like all other internal body structures, poorly supplied with the end organs of touch."[29] This is simply not true. While the innervation of the vaginal walls and the cervix is much less dense than that of the external clitoris, significant sensory information is conveyed from those regions. Selective stimulation of the vaginal walls and the cervix (as well as the labia, perineum, anus, and rectum) can produce clearly detectable sensations as well as localized activation in the somatosensory cortex of the brain (figure 4.4). Furthermore, we now have a better understanding of the anatomy of the clitoris. The external clitoris, a portion of the clitoral glans, is merely the "tip of the iceberg." The deeper portions of the clitoris have a wishbone shape, comprising two bulbs that encompass the urethra and the anterior wall of the vagina (figure 4.6). As a result, stimulation of the anterior vaginal wall (the belly side) can activate sensory nerve endings in the bulbs of the clitoris.[30]

Perhaps the strongest evidence for a purely vaginal orgasm that does not involve the clitoris comes from women with complete spinal cord transection. For these women, touch signals from the clitoris (both the external and the deep portions), carried by the pudendal nerve, cannot reach the brain and therefore cannot be perceived. Some of them can achieve orgasm (verified by brain scanning) from stimulation of the cervix,[8] presumably due to the spared vagus nerve pathway to the brain (figure 4.3). This finding argues strongly for the existence of sexual sensation from the cervix, but does not answer the question of the role of this sensation in the orgasms of uninjured women.

While the existence of the vaginal orgasm as a unique nonclitoral physiological phenomenon continues to be a source of debate,[31] several points are now clear. The clitoris has a special role in triggering orgasm via the pudendal nerve, and it may be achieved most reliably through stimulation of the external clitoris, where the density of sensory innervation is the highest. Penetration of the vagina, as occurs during dildo play, manual stimulation, or penis-vagina sex, can activate touch sensors in the base of the clitoris through the vaginal wall. It can also activate touch sensors in the vaginal wall itself, as well as those in the cervix, and these signals are conveyed to the spinal cord through different sensory nerves: the pelvic, hypogastric, and vagus. Those nerves also carry signals from perigenital areas like the perineum, rectum, and anus. Touch signals from *all* of these sources can contribute to orgasm.

Individual women report a great deal of variation in their experience of orgasm. For some, all orgasms feel more or less the same, while others report qualitatively different types of orgasmic sensations (focused versus spreading, convulsive versus vibrating), depending upon the details of sexual stimulation (with and without penetration, or with and without concomitant breast stimulation, for example). And, of course, cognitive and experiential factors also play an important role. I knew a woman in college who claimed that she could have an orgasm with a man only if he was wearing a singlet made of fishnet fabric. It's unlikely that the basis for this particular erotic requirement lay in the structure of the nerve endings in her skin. A brief perusal of the Internet will yield a nearly infinite variation of requirements for orgasm in both men and women involving fantasy, role play, costumes, etc., most of them unrelated to touch sensation at all.

~~~~~~~

Generally speaking, men report more reliable orgasms than women during both masturbation and partnered sex. This has led to the popular idea that, while women experience a complex blend of sexual sensations from different pelvic areas, the entirety of men's sexual sensation is unvarying and originates from the penis. This is incorrect. Just as the external clitoris has a special role in women's orgasm, the penis, particularly the glabrous skin of the glans penis, is the most reliable trigger of men's orgasm, but other pelvic regions also contribute. After complete removal of the prostate, in the best cases erection, urinary control, and orgasm are spared but ejaculation is eliminated, resulting in dry orgasms.[32] In one recent small survey, all the men who had undergone complete prostatectomy but could still have orgasms reported that they no longer felt the exquisite sensation of inevitability, the "point of no return" leading to orgasm.[33] Signals from the scrotum, testicles, perineum, anus, and rectum can also contribute to male orgasm and can influence both its timing and qualitative aspects (deep versus surface, throbbing versus flowing).[34]

~~~~~~~

The brain has an evolutionarily ancient dedicated circuit for pleasure. It's a complicated one, but the core of it involves activation of a region in the midbrain called the ventral tegmental area and the consequent release of the neurotransmitter dopamine in a number of brain areas that are the target of ventral tegmental area axons, most notably the nucleus accumbens and the dorsal striatum. When we're hungry, eating delicious food activates this dopamine-using pleasure circuit; it is also activated when we have an orgasm. This makes sense: The pleasure circuit is engaged by behaviors that keep us alive and get our genes into the next generation, like eating food, drinking water, and having sexual intercourse. Many psychoactive drugs, including alcohol, nicotine, cocaine, amphetamines, and heroin, can artificially activate this brain pleasure circuit.

We know, however, that the experience of orgasm is more complex than just a feeling of pure pleasure. To help us understand the subjective experience of sexual bliss, it's useful to examine the complete pattern of brain activity during orgasm. In relevant experiments men and women have their heads placed in a brain-scanning machine. While the scientists in the room stand around with their clipboards, trying not to make everyone feel too uncomfortable, the subjects are brought to orgasm by manual stimulation from their partners. In this way, one can compare three conditions: unstimulated, genital stimulation before orgasm, and genital stimulation during orgasm.[35]

The pattern of brain activation during genital stimulation and orgasm is mostly similar in men and women. When the genitals are stimulated but orgasm has yet to occur, patches of activation are seen in the primary somatosensory cortex: one in the contiguous body map and another beyond the toe representation in the mesial postcentral gyrus. Additional patches of activation are seen in the corresponding parts of the body map of higher somatosensory processing areas, such as the secondary somatosensory cortex. As genital stimulation continues, one sees a reduction in the ongoing activity in the amygdala, a region that processes emotional signals related to fear.[36] The interpretation of this finding is that continued genital stimulation is associated with a reduction of fear and a consequent loss of vigilance toward potential threats—in other words, relaxation.

When orgasm occurs, in addition to activation of the somatosensory cortex and deactivation of the amygdala, we also see activation of key portions of the pleasure circuit: the ventral tegmental area, the nucleus accumbens, and the dorsal striatum. Another activated region is the cerebellar nuclei, an area involved with motor coordination as well as certain aspects of emotional cognition. It's possible that cerebellar activation is related to the involuntary nature of thrusting and writhing movements as well as involuntary facial expressions during orgasm. Certain brain regions are also deactivated during orgasm. These include the lateral orbitofrontal cortex and the anterior temporal pole, areas involved in thoughtful decision making, self-control, moral choice, and social evaluation. Clearly, the moment of orgasm is not the time you want to be making careful decisions about your life, as many of the most important brain regions for this function have temporarily gone offline.

Most brain-scanning machines have poor temporal resolution: They must average activity that occurs over the course of several seconds. As a consequence we don't have a detailed sense of the temporal progression of brain-region activation or deactivation during genital stimulation and orgasm. Presumably, stimulation of touch sensors in the genitals leads to activation of the genital areas on the somatosensory map. Beyond that, the details are unclear. It could be that the genital areas of the somatosensory cortex, when repeatedly stimulated in the right conditions, send signals to activate regions like the ventral tegmental area and deactivate others like the amygdala and lateral orbitofrontal cortex. The flow of signals between brain regions during orgasm remains poorly understood.

~~~~~~

Here's the miracle: When we have an orgasm it feels like a transcendent, unified moment, not merely a collection of disparate sensations. We experience orgasms as intrinsically pleasurable and emotionally positive. Why is this? If one were simply to sum up the brain scanner results, we'd get a sort of recipe.

For an orgasm, mix together the following ingredients:

Activation of touch sense from the genitals (somatosensory cortex)
Deactivation of fear and vigilance (amygdala)
Activation of the pleasure circuit (ventral tegmental area, nucleus accumbens, dorsal striatum)
Activation of a motor control center (cerebellar nuclei)

Deactivation of areas involved in slow decision making
(lateral orbitofrontal cortex and anterior temporal pole)

Serves: 1

If this is really the recipe for an orgasm, then we can imagine a fanciful thought experiment: Could artificial activation and deactivation of this constellation of brain regions result in an experience that effectively simulates natural orgasm? We don't know, and the closest thing we do have to such a scenario is a natural experiment: In some people, epileptic seizures can produce an orgasm.[37] It's unlikely that any of these seizures produce the particular pattern of activation and deactivation included in our orgasm recipe, and so it is not surprising that the patients who experience them point out the differences between their seizure-evoked orgasms and those resulting from sexual activity. For example, it's unlikely that the vigilance and cognitive effects produced by deactivation of the amygdala and decision-making cortical areas are found in seizure-evoked orgasms.

Seizure-evoked sexual sensations are not all identical. When seizures are restricted to the parietal lobe, they can produce the sensation of touch to the genitals and perigenital areas, but they do not result in the intense pleasurable sensation that is the hallmark of orgasm. When seizures are restricted to the midbrain/temporal lobe region, they can result in a wave of pleasure, but that pleasure has no particular genital or even tactile focus. It is more akin to the diffuse rush of pleasure one gets from opiate drugs. It is only when both the somatosensory cortex in the parietal lobe and the pleasure circuit in the midbrain/temporal lobe are activated together that a seizure can produce a sensation that approaches that of a natural orgasm.

Although we experience orgasms as intrinsically pleasurable, they are really just a trick our brains are playing on us through simultaneous activity in many brain regions. The gestalt feeling of an orgasm is divisible into its component parts. The discriminative sense of genital touch is mediated by activation of the somatosensory cortex, but that is not pleasurable in and of itself. It has no emotional valence. That pleasurable emotional feeling is produced only when the ventral tegmental area dopamine neurons are activated. And the accompanying relaxation (loss of vigilance) and cognitive disinhibition during orgasm result from their own regional deactivation effects. The great thing is that we don't have to think about all these disparate ingredients that go into producing an orgasm. We just get to enjoy the unified sensation that the brain cooks up.

CHAPTER FIVE

HOT CHILI PEPPERS, COOL MINT, AND VAMPIRE BATS

ere's the plan: I'm going to give you a backpack filled with Ziploc bags—some will be filled with fresh mint leaves and others with juicy habanero chili peppers. You'll also get a clipboard, a pencil, a spare pair of socks, and a round-the-world airplane ticket. Your job, should you choose to accept it, is to travel around the world and visit all sorts of places, from the biggest cities to the most remote jungle encampments. In each location you will seek out a wide variety of people—young and old, rich and poor—and then rub chopped-up mint leaves or diced chili peppers on their skin, ask them to describe the sensation, and record their responses. Try applying the samples on both the glabrous skin of the lips and the hairy skin of the forearm. (These substances don't have to touch the tongue for their effects to be experienced.)

If you conducted this survey where I live, in Baltimore, you'd find that the dominant word used to describe the tactile experience of the habanero, smeared on either the lips or the forearm, would be *hot*, while for mint it would be *cool*. Is this merely a convenient turn of phrase, a colloquialism? After all, if we were to use a thermometer to measure the actual temperature of mint or chili peppers, we'd find that they are not literally hot or cool. And Baltimoreans (like many others) often use these words metaphorically—to mean, for example, stylish ("the Tesla Roadster looks so cool") or sexually attractive ("Rachel Weisz is so hot"). The use of words like *cool* to mean "stylish" and *hot* for "sexually attractive" are metaphors that are specific to a particular time and place. People in Shakespeare's time, for example, appear not to have used either of these linguistic constructions. Are "hot chili peppers" or "cool mint" also local, culturally constructed metaphors, or do they reflect some deeper biological reality? If they are merely cultural constructions, you would expect to find groups of people in your world-traveling survey who don't describe the tactile sensation of chili peppers as hot or mint as cool.

If, however, your survey did indeed reveal that these figures of speech are widespread, would that constitute proof that hot chilies and cool mint are biologically determined metaphors? Not exactly. To play devil's advocate, one could imagine that, over many years with widespread communication, the idea of hot chili peppers and cool mint originated in one place and spread around the world through cultural contact. While various species of mint are widely dispersed geographically, chili peppers originated in South America and had been carried only to Central America and the Caribbean before European colonization. They were unknown in Europe, Africa, or Asia before Columbus returned from the New World. Soon after, they were spread by European powers, notably Spain and Portugal, to their other colonies. It's hard to imagine now, but the foods of places like India and Thailand had no fiery chili pepper before the sixteenth century.[1] At present it's unclear if there are any places left on earth where chili peppers have not been introduced.[2] So, to really do this survey properly, you'd also need a time machine to take you back to, say, Thailand in the fifteenth century and do your mint/chili survey there as well.

To my knowledge, this type of ethnographic (not to mention time-travel) survey has yet to be done, but from a biological perspective we can predict how it would turn out. Given what we know about the biology of touch, we'd predict that nearly every person around the world would describe chili peppers as hot and mint as cool, even if he or she were experiencing these tactile sensations for the first time and had never heard others describe them. It appears that the cool-mint and hot-chili-pepper metaphors are biologically hardwired from birth.

The main active ingredient in mint is menthol, while its equivalent in chili peppers is a chemical called capsaicin. Less potent chili peppers, like the Anaheim, have a low concentration of capsaicin, while very strong ones, like the Bhut Jolokia pepper, can produce about one-thousand-fold more.[3] So why *are* we biologically predisposed to perceive menthol as cool and capsaicin as hot? One possibility is that there's a class of nerve ending in the skin that can sense cooling and a different class that can respond to menthol. The signals conveyed by these distinct fibers could then ultimately converge in the brain: Mint and cooling might feel the same because they activate the same brain region dedicated to the sensation of cooling. In an analogous fashion, separate heat-sensing and capsaicin-sensing nerve fibers could ultimately send their impulses to a heat-sensitive brain region.

This hypothesis, therefore, rests on signal convergence in the somatosensory cortex, and while it's reasonable and appealing, it's actually dead wrong. How do we know that? First, we can record electrical signals from single sensory nerve fibers in the arm that respond to both heat and capsaicin, and other single nerve fibers that respond to both menthol and cooling. These show that temperature and chemical signals are present in the neurons that innervate the skin long before any signals reach the brain. We also have some molecular evidence. There are free nerve endings in the epidermal layer of the skin (figure 2.3) that contain a sensor on their outer membrane called TRPV1. This single protein molecule can respond to both heat and capsaicin by opening an ion channel, a pore that lets positive ions flow inside, thereby causing the sensory neuron to fire electrical spikes. Similarly, there are free nerve endings that contain a different sensor, called TRPM8, that can respond to both menthol and cooling. The answer to our puzzle is that the metaphor is not in the culture, or even in the brain region. The metaphor is encoded within the sensor molecules in the nerve endings of the skin.

How did this molecular metaphor develop? How did thermosensors like TRPV1 and TRPM8 become sensitive to plant products like capsaicin and menthol? We can't know for certain the sequence of evolutionary events that gave rise to these two dual-function sensors. The best guess is that TRPV1 and TRPM8 evolved in some animals as temperature sensors and that certain plants later developed compounds that would activate them in order to deter their consumption by predators. Plants that produced menthol and capsaicin would therefore have a survival and reproductive advantage and become more prevalent in the population of that species. In this scenario it's plant evolution that initially drove the dual-function properties of the sensors, not animal evolution.

~~~~~

David Julius and his coworkers at the University of California, San Francisco, have studied the molecular properties of TRPV1 and TRPM8 by using genetic tricks to force kidney cells or frog eggs grown in a culture dish to produce great quantities of TRPV1 or TRPM8 while recording the electrical signals that pass across the cell membrane when these sensors are stimulated.[4] These studies have revealed that features of these molecules explain aspects of our everyday tactile experiences. For example, the oil of the eucalyptus tree contains a substance called eucalyptol that, like menthol, can activate TRPM8 to produce a cooling sensation. This is why eucalyptus extract is often used in soothing skin creams, mouthwash, and throat lozenges.[5]

TRP function can also have an impact on our experience of a summer's day at the beach. If you're out in the sun too long, the resulting sunburn will set in motion a cascade of inflammatory processes in your skin, including the production of compounds called prostanoids and bradykinin. These chemicals have the property of reducing the temperature threshold of TRPV1 activation from 109°F to 85°F. As a consequence, when you return home from the beach and step in the shower to rinse off the remaining sand and sunscreen, the water temperature you typically select will now be too hot, and you'll have to reduce it to avoid a painful burning sensation.[6]

Another example involves the bird feeder in your backyard. While mammals have the standard form of TRPV1, activated by both capsaicin and heat, birds are utterly indifferent to capsaicin, as they can't detect it at all. (Birdwatchers often spike the seeds in their feeders with chili peppers to deter squirrels, raccoons, and other mammals while leaving the birds unaffected.) When the TRPV1 gene is extracted from a bird and expressed artificially in kidney cells, it reveals a bird-variant form of TRPV1 that responds to heat but not capsaicin. Examination of the sequence of bird DNA can pinpoint the change to the exact spot that's necessary for capsaicin binding, located on the inner surface of the cell's outer membrane.[7] Interestingly, chili pepper plants and birds appear to have reached a satisfying sort of evolutionary détente. When mammals eat chili peppers, they tend to destroy the seeds with their molars. Birds, on the other hand, don't have molars and so pass most of the seeds through their digestive system intact. When they defecate, they spread viable chili pepper seeds to new locations. It's a win-win situation for birds and chilies.[8]

~~~~~

A few years after the initial identification of TRPV1, several groups used genetic engineering techniques to produce mice that lacked TRPV1 and measured their responses to capsaicin and heat. These mutant mice were found to completely lack behavioral and electrical responses to capsaicin. However, their responses to heat were diminished but not completely eliminated. For example, when their tails were placed in hot water (122°F), they eventually flicked them away, but it took four times longer than for normal mice. Likewise, the ability of inflammation to boost heat sensation was reduced but not eliminated in the mutant mice.[9] These results indicate that there must be other heat sensors in addition to TRPV1.

Initially it seemed as if a satisfying explanation was in hand when a family of TRPV channels was identified with a range of heat sensitivities: TRPV4 and TRPV3, when expressed in kidney cells in a culture dish, responded to warm temperatures below the range of TRPV1. Conversely, TRPV2 responded to extreme heat (>125°F), well above the threshold for TRPV1. In this way, successive activation of various TRPV channels with different thresholds could potentially detect a range of skin temperatures encountered in real life, from tepid to warm to hot to painfully hot (figure 5.1). In addition to being expressed in free sensory nerve endings, TRPV3 and TRPV4 were also found in keratinocytes, the main cell type of the epidermis where the free nerve endings terminate. This suggested that the neighboring skin cells might play a role in helping the free nerve endings to detect gentle warmth. TRPV3, one of the gentle-warmth detectors, was also shown to be activated by compounds from a wide range of spices, including camphor, nutmeg, cinnamon, oregano, cloves, and bay leaves, some of which are associated with a perception of warmth. (As a child I was an enthusiastic consumer of Red Hots, a cinnamon-flavored candy.)

The prediction from figure 5.1 is clear: TRPV4 and TRPV3 should be required for detecting gentle warmth, and TRPV2 for extreme heat. Taken together, these three additional TRPV sensors should account for the residual heat perception present when the TRPV1 gene is deleted or the TRPV1 protein is blocked by a drug. Surprisingly, when mutant mice were created that lacked TRPV3, TRPV4, or TRPV2, either alone or in combination, they showed no significant deficit in heat perception in a wide variety of tasks. This result strongly suggests that there are even more heat detectors in the skin

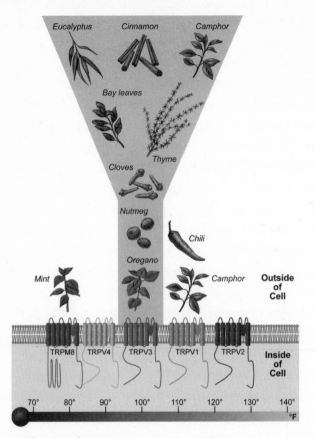

Figure 5.1 A family of temperature-sensitive TRP sensors can respond to heating, cooling, and various pungent chemicals found in plants. Here, each TRP sensor is shown at a position along the thermometer where it begins to respond to heating or cooling relative to skin temperature.

Keep in mind that while core body temperature is about 99°F, temperature in the epidermal layer of the skin is about 90°F. While the TRP sensors are not identical, they share certain properties: All of them thread their way across the cell membrane six times and all have a loop structure that dips into the membrane to form an ion channel. Each TRP sensor is drawn to show its overall molecular structure. It is worth noting that the thermal activation points for all the TRP sensors are not sharp thresholds—there is quite a bit of cell-to-cell variation. Adapted from L. Vay, C. Gu, and P. A. McNaughton, "The thermo-TRP ion channel family: properties and therapeutic implications," *British Journal of Pharmacology* 165 (2012): 787–801, with permission of the publisher, Wiley.

that we have yet to identify, and that these are likely to be molecules that are not part of the TRPV family of genes.[10]

A similarly murky situation surrounds the sensation of cooling. When genetic engineering was used to create mice that lacked TRPM8, they showed a complete loss of responses to menthol and eucalyptol applied to the skin and an incomplete reduction in responses to mild cooling. In particular, their responses to gentle cooling (below 77°F) were profoundly diminished, but their responses to severe cold (less than 58°F) were normal.[11] Similar to the partial effect of TRPV1 deletion of heat sensing, this result indicates that there must be additional molecular sensors for cold, particularly severe cold, that remain undiscovered.[12]

~~~~~~

When rubbed on the skin, mint feels cool and chili peppers feel hot, but what is the sensation produced by horseradish, or its Japanese cousin wasabi? It's not exactly hot, but more like a warm pungency. Wasabi, horseradish, and yellow mustard all contain a chemical called AITC (allyl isothiocyanate), which activates a different sensor of the TRP family called TRPA1.[13] Another TRPA1-activating group of compounds includes allicin and DADS (diallyl disulfide), which are found in garlic and onions and account for their effects on skin sensation, including the eye-watering response evoked by activation of TRPA1 in the cornea.[14]

When I lived in Chicago in the 1980s, there was a great Italian bar on Halsted Street that served steamed garlic that you could smear on crusty bread and wash down with Moretti beer. The dish was prepared by gently removing some of the outer papery skin and then steaming the whole garlic bulb intact. Only after it was completely cooked would the chef cut it in half, around the equator, to allow the diner to scoop out the mild soft flesh of the plant from each clove with a special tiny knife. What chefs have known for years is that the pungent chemicals in garlic and onions—the ones that cause irritation of the skin and eyes—are produced only

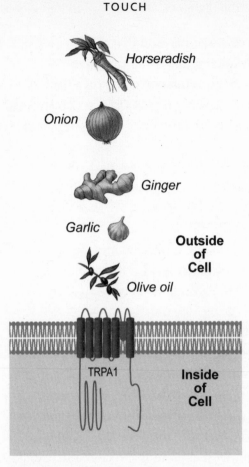

Figure 5.2 The TRPA1 sensor, whimsically called the wasabi receptor, is activated by a wide variety of pungent compounds from plants, most notably wasabi, horseradish, and yellow mustard, as well as structurally similar products from onions and garlic and a structurally distinct compound, oleocanthol, found in extra-virgin olive oil. It's interesting that several different families of plants, most notably the wasabi/horseradish/mustard family (called *Brassicaceae*) and the onion/garlic/leek/shallot family (called *Allioideae*) have independently evolved chemicals to activate TRPA1, presumably to reduce predation, although these compounds also have antimicrobial properties. Adapted from L. Vay, C. Gu, and P. A. McNaughton, "The thermo-TRP ion channel family: properties and therapeutic implications," *British Journal of Pharmacology* 165 (2012): 787–801, with permission of the publisher, Wiley.

when the bulb is cut or crushed. When the bulb is intact, the enzyme that produces allicin and related pungent compounds is trapped within special compartments inside the plant cells and cannot act on its substrate. Allicin and DADS are also partially degraded by the high temperatures of cooking. This means that cooking an intact onion or garlic bulb will produce a low concentration of TRPA1-activating pungent compounds, little skin and eye irritation, and a delicious, mild appetizer.[15, 16]

~~~~~

The prickly ash tree, *Xanthoxylum*, is also known as the tickle-tongue tree or the toothache tree because its sap or berries produce a numbing, tingling sensation when ingested. Indeed, the berries of *Xanthoxylum*, also called Szechuan peppercorns, are prized for the tingling sensation they add to spicy dishes from this region of China. These tingles suggest an interaction with sensory neurons. In both East Asia and North America, preparations of prickly ash are used as folk medicine for their anesthetic or pain-masking properties. The active ingredient of *Xanthoxylum* is a chemical called hydroxyl-alpha-sanshool. Considering what we have learned about the actions of other plant compounds on sensory neurons that innervate the skin, an obvious guess would be that hydroxyl-alpha-sanshool activates some type of TRP channel in these cells. This, however, is not the case. Hydroxyl-alpha-sanshool excites sensory neurons through a novel mechanism by blocking an ion channel called the two-pore potassium channel. This type of channel normally allows the slow leak of positive ions out of the neuron, so that when it is blocked, positive charge builds up quickly inside the cell, ultimately causing it to fire spikes and thereby send signals to the brain. The neurons that are activated by preparations of *Xanthoxylum* include C-tactile fibers, which convey light pleasant touch; the caress sensors; and the Meissner fibers, which convey vibration at moderate frequencies. It's not entirely clear why their activation produces a tingling sensation.[17]

~~~~~

Vampire bats are amazing, and not only for their fanciful roles in horror films.[18] We've discussed them briefly earlier in the context of their social grooming to solicit a shared blood meal (chapter 1), but now let's examine their tactile specializations for feeding. Vampire bats have a unique ecological niche: They are the only known mammals whose entire food supply consists of blood from warm-blooded animals (other mammals and birds). Some species of bat will eat insects or fruits, but vampire bats can only swallow liquids and will starve to death before they will consume a nonblood meal. Vampire bats fly to select target prey and typically alight on their backs or the crests of the necks. They then proceed to search for a suitable spot to carefully bite and extract about two teaspoons of blood. They search for a place that's not encumbered by too much hair or fur and where blood vessels run close to the surface of the skin. This search for a buried blood vessel is the moment when the ability to detect heat at a distance is particularly useful. Ludwig Kürten and Uwe Schmidt of the University of Bonn have brought vampire bats into the lab and shown that they can detect the infra-red radiation emitted from living human skin at a distance of about 6 inches.[19]

Many species of bat have a facial structure called a nose leaf, which is thought to aid in the echolocation of prey, but only vampire bats have a set of three nasal pits surrounding the nose leaf (figure 5.3). The skin of these pits is thin, hairless, and devoid of glands, making it an ideal location to house infrared sensors. The pits are also separated from the surrounding parts of the face by a layer of dense connective tissue that serves as a thermal insulator. As a result, the temperature of the nasal pits is about 84°F, substantially cooler than the 99°F temperature of the surrounding skin. This allows the heat sensors in the nasal pits to distinguish between the heat of prey and the heat of the bat's own face.

So what sensor does the vampire bat use to detect infrared

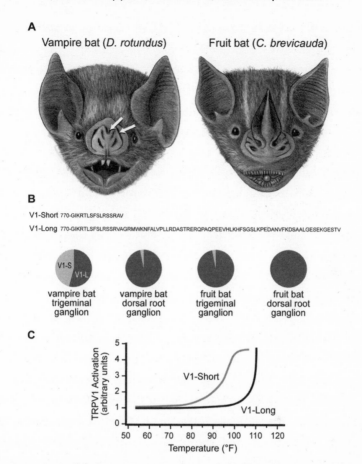

A

Vampire bat (*D. rotundus*)   Fruit bat (*C. brevicauda*)

B

V1-Short 770-GIKRTLSFSLRSSRAV

V1-Long 770-GIKRTLSFSLRSSRVAGRMWKNFALVPLLRDASTRERQPAQPEEVHLKHFSGSLKPEDANVFKDSAALGESEKGESTV

V1-S
V1-L

vampire bat   vampire bat   fruit bat   fruit bat
trigeminal   dorsal root   trigeminal   dorsal root
ganglion   ganglion   ganglion   ganglion

C

TRPV1 Activation (arbitrary units)

V1-Short

V1-Long

Temperature (°F)

Figure 5.3 A modified supersensitive form of TRPV1 allows vampire bats to detect infrared radiation. (A) Vampire bats, which can detect infrared radiation, have nasal pits, indicated by arrows, while fruit bats, which cannot detect infrared radiation, do not. (B) The amino acid sequence of the carboxy-tail-end of two alternatively spliced forms of TRPV1, the super-heat-sensitive short form and the normally heat-sensitive long form. The short form is highly expressed in neurons from the trigeminal ganglion that innervate the face (including the nasal pits) but not in the dorsal root ganglion neurons that innervate the body of the vampire bat. (C) When artificially expressed in kidney cells grown in a dish, the supersensitive short form of TRPV1 begins to be activated at 86°F while the long form is activated only at temperatures above 109°F. Adapted from E. O. Gracheva, J. F. Cordero-Morales, J. A. Gonzales-Carcacia, N. T. Ingolia, C. Manno, C. I. Aranguren, J. S. Weissman, and D. Julius, "Ganglion-specific splicing of TRPV1 underlies infrared sensation in vampire bats," *Nature* 476 (2011): 88–91, with permission of Nature Publishing Group.

radiation? We already know that TRPV1 in humans and mice can detect temperatures greater than 109°F, but clearly that's insufficiently sensitive. To identify the infrared sensor in vampire bats, David Julius, Elena Gracheva, and their colleagues performed a clever experiment. They gathered vampire bats and fruit bats (which can't sense infrared radiation). Then they carefully dissected out the clusters of neuronal cell bodies that innervate the face (the trigeminal ganglion) and analyzed their expression of the TRPV1 gene. They found that there are actually two different forms of TRPV1 expressed in trigeminal ganglion sensory neurons: a long form, which has the conventional heat threshold of 109°F, and a short form, which is activated at a much lower temperature, about 86°F, just above the resting temperature of the nasal pits. Fruit bats have only the long form in their trigeminal ganglia, while vampire bats have both, in roughly equal measure.[20]

The lovely result is that the vampire bat has evolved a supersensitive form of TRPV1 to enable it to detect infrared radiation for feeding. But what does this mean for the rest of its body? After all, the vampire bat needs to detect heat in other body parts as well. When dorsal root ganglia—clusters of neurons that innervate non-face areas—were examined, they showed only trace amounts of the supersensitive short form of TRPV1. This explains why vampire bats can maintain normal thermal sensitivity in other body regions that are not used for locating a blood meal.

~~~~~~

I'm sure that you've been lying awake pondering this question: If you blindfold a rattlesnake, can it still accurately strike its prey? Thanks to a group of intrepid researchers led by Peter Hartline at the University of Illinois at Urbana-Champaign, we know the answer. These investigators (carefully) blindfolded their rattlesnakes and placed them on a pedestal at the center of a circular enclosure. Then they induced them to strike by jiggling a heat source (the hot tip of a soldering iron) in an enticing fashion to mimic the movements of a warm-blooded animal. The soldering iron was placed at various angles in the direction the snake was facing and just outside of striking range (about three feet away). Even with both eyes completely covered, the snake was able to strike accurately, within five degrees of the target. As the authors of this study noted, "This is very impressive, and for a mouse it is deadly."[21]

How does the rattlesnake accomplish this? It's not by the sense of smell: The snakes will strike accurately at a warm, odorless object or one that is completely encased in an odor-blocking shield. However, if you place the warm object behind a special pane of glass that blocks infrared radiation, it can no longer strike accurately. Like vampire bats, rattlesnakes can sense the infrared radiation emitted from warm objects. Rattlesnakes, however, are much more sensitive and are able to detect warm objects at a maximum distance of about 39 inches, compared to 6 inches for vampire bats. The structure that confers the infrared sensitivity of the rattlesnake is the pit organ, a small cavity located between the eye and the nostril (figure 5.4). If the pit organs on each side are covered or damaged, then the rattlesnakes can no longer strike accurately when they are blindfolded or placed in the dark. Rattlesnakes are not the only type of snake with an infrared-detecting pit organ. They are one species of a group of related snakes called pit vipers (subfamily *Crotalinae*), which include moccasins, lanceheads, and bushmasters in the Americas and temple vipers and hundred-pace vipers in Asia.

The pit organ functions like a crude pinhole camera. There is a small aperture at the front, and a thin infrared-sensitive membrane at the rear, stretched so that there is an air space on either side of it. Figure 5.4B shows how the aperture of the pit organ restricts infrared radiation so that a source at a particular point in space will strike only a small section of the pit membrane, enabling the pit organ to form a low-resolution picture of the infrared world. The pit membrane is innervated by about seven thousand sensory fibers from the snake's trigeminal ganglion that then carry information encoding the rattlesnake's infrared map of the world to a part of

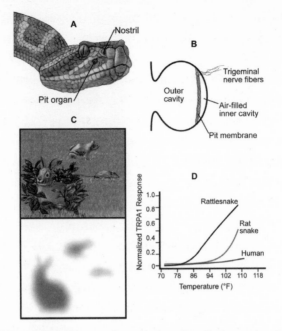

Figure 5.4 The rattlesnake can detect infrared radiation using its pit organ, which contains a modified temperature-sensitive form of TRPA1. (A) The pit organ is located between the eye and the nostril. (B) A cross-section of the pit organ shows how it functions as a crude pinhole camera to localize prey. Nerve fibers from cells in the trigeminal ganglion branch in the pit membrane, which is stretched like a drumhead over an air-filled apace. (C) An artist's rendering of the visual (top) and infrared (bottom) sensory worlds of the rattlesnake. These two streams of information are aligned and combined in the snake's brain. Note that the snake can detect the vague outline of the warm rabbit with infrared sense even when it is hidden in a bush (or it is dark outside). The snake's infrared sense can be used to detect not only warm objects against a cooler background but also cool objects against a warmer background, like the frog emerging from a pond onto sun-warmed grass. (D) TRPA1 from a rattlesnake is genetically modified so that it can be activated at temperatures above 86°F. TRPA1 from a rat snake, a species without infrared sense, is only weakly activated by warming, and human TRPA1 is not activated at all. Panels A, B, and D are adapted from E. O. Gracheva, N. T. Ingolia, Y. M. Kelly, J. F. Cordero-Morales, G. Hollopeter, A. T. Chesler, E. E. Sánchez, J. C. Perez, J. S. Weissman, and D. Julius, "Molecular basis of infrared detection by snakes," *Nature* 464 (2010): 1006–11, with permission of Nature Publishing Group. Panel C is adapted from E. A. Newman and P. H. Hartline, "The infrared 'vision' of snakes," *Scientific American* 246 (1982): 116–27, with permission of Macmillan Publishers.

the brain called the optic tectum, where it is combined with visual information in a fashion that aligns the visual and infrared maps (figure 5.4C).[22]

One might guess that the molecular sensor that detects infrared radiation in the pit organ of the rattlesnake is the same supersensitive form of TRPV1 that's used by vampire bats. However, when David Julius and his colleagues examined the trigeminal ganglia (where the sensory neurons that innervate the pit organ reside), they found that TRPV1 was not enriched there, as one would expect if it were the pit-organ infrared sensor. However, they surprisingly did discover that the wasabi receptor, TRPA1, was elevated four-hundred-fold in the snake's trigeminal ganglion. This was an odd finding, because mammalian TRPA1 is not activated by heat at all. When human and rattlesnake TRPA1 were expressed in kidney cells, heat was shown to activate rattlesnake TRPA1 at temperatures above 86°F; however, human TRPA1 was almost entirely insensitive to heat. Rat snakes, which do not possess facial infrared-sensing organs, have a form of TRPA1 that is only weakly heat-sensitive.[23] If we have come to think of TRPA1 as the wasabi sensor, it is only because we happened to study mammalian TRPA1 first. If we had a more vipercentric world-view, we'd say that TRPA1 is a heat sensor that can also be activated by wasabi and garlic.

Boas and pythons are from a snake lineage that is approximately 30 million years older than the pit vipers. They also have infrared-sensing pits, typically thirteen per side, located in two rows, one above and one below the mouth. The openings of these pits are not constricted and so they do not function like pinhole cameras. Rather, each pit has a slightly different field of view based upon its position on the snake's face. From behavioral tests we know that pythons and boas are not as sensitive to infrared radiation as rattlesnakes are, so it was not entirely surprising to find that TRPA1 from pythons was less heat-sensitive than that of a rattlesnake but more sensitive than rat snake TRPA1. When the sequences of the TRPA1 genes for humans, pythons, and rattlesnakes are compared, one can see that modification of TRPA1 to render it heat-sensitive evolved twice in the snake lineage: once in the ancient boas and pythons, and then again in the more modern pit vipers.[24] Sometimes the process of random mutation and natural selection will yield a related molecular and structural solution to a problem (like infrared sensing) in different organisms, millions of years apart. This is the wonderful process of convergent evolution.

Not all creatures use their infrared detectors to find prey. For example, most animals run or fly away from forest fires, but the fire beetles of North America (called *Melanophila*) are drawn toward them. It's not a desire for self-immolation that compels them, however. The beetles arrive at the site of a fire just as the flames have subsided and then copulate in the comfortably still-warm ashes. The female then deposits her eggs under the charred bark of newly burned pine trees. When the fire beetle larvae hatch the following summer, they can feed on the charred wood. (Living wood has chemical defenses that make it inedible to the larvae.) In some cases fire beetles have been drawn to other hot sites, including factories and even a football game held in a stadium where lots of the spectators were smoking cigarettes. Perhaps the most dramatic infestation of this type occurred in the Central Valley of California in August 1925. When a huge fire consumed an oil tank near the town of Coalinga, huge numbers of fire beetles began to swarm toward it. Newspaper reports of the time estimated that millions of fire beetles converged on Coalinga and remained for several days after the fire was extinguished.

Because Coalinga is situated in an arid valley, the best guess is that these beetles came from a site in the western foothills of the Sierra Nevada mountains, approximately eighty miles away. *Melanophila* beetles have a single infrared-detecting pit on each side of the abdomen. Many years later, when Helmut Schmitz and Herbert Bousack of the University of Bonn performed calculations to estimate the amount of infrared radiation that would have fallen on these sensors at a distance of eighty miles, they found that it was so small, it was embedded within ambient thermal noise produced by the fire beetle's body. The nervous system of the fire beetle has a difficult engineering task to extract this tiny signal and use it to trigger migratory behavior. To date we do not know if the infrared sensors of the fire beetles use TRPV1 like vampire bats, TRPA1 like rattlesnakes, or an entirely different mechanism—perhaps not even a member of the TRP family at all.[25]

~~~~~~~

If you type the word *paradise* into a Google Images search, the result will be a screen filled with one hundred different images, each one of which will be a view of a tropical beach. What's the explanation for that? In part—at least for people living in affluent societies—a tropical beach suggests a leisurely vacation. But why, then, doesn't the search term *paradise* bring up pictures of other popular vacation spots, like New York City, or a ski resort, or Disneyland? The reason is the weather: Paradise is a place where our bodies don't have to work very hard to maintain our core temperature of approximately 99°F. We humans and other homeothermic animals (mammals and birds) cannot tolerate deviations of our core temperature of more than a few degrees. If it's hot, we engage in both reflexive and voluntary activities: We sweat, vasodilate, have a cold drink, or jump in a swimming pool to cool our core. If it's cold, we shiver, vasoconstrict, and put on a sweater. These homeostatic reflexes and behaviors require that we constantly monitor our internal core temperature and the temperature of the outside world as sensed through the skin. We need to know when our skin is cold or hot enough to require a physiological response to maintain our core temperature within a narrow range. The thresholds of human TRPM8 and TRPV1 are well calibrated for this task: TRPM8 is activated at temperatures below 78°F, and TRPV1 is activated at temperatures above 109°F.

If the TRPM8 and TRPV1 activation thresholds for detecting cooling and warming are truly designed to help us maintain our core temperatures, then we should expect these thresholds to be different in animals with core temperatures unlike our own. Indeed, when DNA encoding TRPM8 from a chicken, a rat, and a frog (the clawed frog *Xenopus laevis*) was used to artificially express TRPM8 channels, it was shown that chicken TRPM8 was tuned to have a warmer threshold of activation: around 86°F, befitting defense of its core temperature of 107°F. The frog, which is not homeothermic and hence needs only to sense extreme cold, has its TRPM8 tuned cooler,

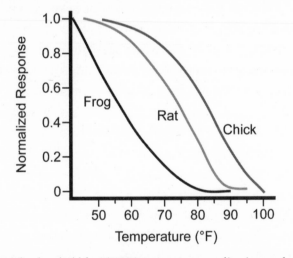

Figure 5.5 The threshold for TRPM8 response to cooling is correlated with core body temperature. These temperature response curves were generated through artificial expression of TRPM8 from a frog, a rat, and a chicken. Adapted from B. R. Myers, Y. M. Sigal, and D. Julius, "Evolution of thermal response properties in a cold-activated TRP channel," *PLOS One* 4 (2009): e5741, an open-access article distributed under the terms of the Creative Commons Attribution License, which permits unrestricted use, distribution, and reproduction in any medium, provided the original author and source are credited.

activating only at temperatures below 66°F (figure 5.5).[26] On the heat-sensing side, thresholds also appear to be set by core temperature. For example, human TRPV1 is activated at temperatures above 109°F, but the zebra fish, which is not homeothermic, has TRPV1 that activates at about 91°F. The bottom line: The thresholds for hot and cold detection in various animals are not random. Rather, they make sense in terms of the temperature regulation each animal needs to achieve in order to function physiologically.[19]

~~~~~~

Years ago, I went on a family visit to my former sister-in-law's house in Ohio in the late fall. When I awoke in the basement guest room, it seemed more than a little chilly. After bundling up, I headed upstairs and checked the thermostat, which was set to 52°F. That was just right for her, for she was standing happily in the kitchen in shorts and a T-shirt. The following year I made sure to bring along a portable space heater for the guest room. I have another friend who heats his house to the point where the doorframes start to warp, and you almost expect to see the sun-bleached skull of a coyote lying on the hardwood floor.

What accounts for such extremes in individual temperature preference in humans? We know that people with a thinner layer of body fat tend to prefer warmer temperatures, which makes sense in terms of core temperature regulation. We also know that those who are more physically active, even if that activity is mere fidgeting, produce more heat through muscular contraction and thereby prefer cooler temperatures. This may partly explain why young children and teenagers are often resistant to putting on a coat. There is also a daily cyclic variation in core body temperature, and hence external temperature preference. But are there actual differences in the temperature-sensing molecules or circuits of the skin and brain that might explain some of these individual variations? Do temperature preferences run in families? At present, the best answer to these questions is maybe.

We know that different species of animal can have TRPV1 and TRPM8 variants that are tuned to different temperatures. It has also been shown in both rats and humans that drugs that block TRPV1 can produce hyperthermia, a rise in core body temperature, and that drugs that activate TRPM8 can do the same. There is also a rare recessive mutation in humans called WNK1/HSN2. Two copies of this mutant gene produce severe degeneration of sensory neurons, but single-copy carriers of the mutation are unaffected. However, careful measurement of heat and cool detection thresholds showed that the WNK1/HSN2 carriers have their warm threshold shifted slightly cooler and their cold threshold shifted slightly warmer than age- and sex-matched controls.[27] But the simple result that one might expect—that genetic variation in human TRPV1 or TRPM8 could account for a portion of individual temperature preference—has yet to emerge. I strongly suspect that my former sister-in-law has a froglike TRPM8 gene, but that has yet to be confirmed.

CHAPTER SIX

PAIN AND EMOTION

On his fourteenth birthday, eager to perform an especially daring trick to impress his friends, a boy jumped off the roof of his family's house in Lahore, Pakistan. On landing, he got up off the ground and said that he was okay, but died the following day from massive internal bleeding. Despite his major injuries, he never indicated that he was hurt, so his family didn't take him to the doctor. Not surprisingly, this was no ordinary boy. He was well known as a street performer who would insert knives through his arms and stand on burning coals. People in his neighborhood said that he was fearless because he could feel no pain at all.

Although the boy died before he could be carefully examined, subsequent investigations by Geoffrey Woods, a geneticist at Addenbrooke's Hospital in Cambridge, England, identified six additional cases in which the ability to sense pain was entirely absent from birth.[1] All were children from families of the Qureshi birdari clan in rural northern Pakistan, but because the condition is the result of a rare and random genetic mutation, it can occur anywhere in the world.[2] The Pakistani families that Woods studied had emigrated to England and sometimes practiced cousin-marriage. None of the six children had felt any type of pain at any time during their lives—not in their skin, muscles, bones, or viscera. They rarely cried as babies. It's not that they felt pain but were indifferent to it; rather, they did not experience pain at all. Neurological exams revealed that the children had normal touch perception of fine mechanical stimuli (vibration, pressure, texture), gentle warmth and cooling (but not painful extremes of temperature), tickling, and caressing. If they were to accidentally hit themselves on the thumb with a hammer, they would feel the pressure of the blow, but no pain from it. The injured thumb would then swell, but it wouldn't throb or ache.

The deficit of the affected children was remarkably specific to pain perception: They had normal reflexes, unimpaired bowel and bladder control, and no obvious disorders of cognition, mood, or social interaction. When asked to describe the meaning of the word *pain,* none of them could give an appropriate answer, though the older ones had learned what actions would be likely to elicit pain in others (to the point where they could convincingly mimic pain after soccer tackles to draw a penalty for the opposing player). Importantly, the loss of physical pain hadn't diminished the ability of these children to experience emotional pain. Their feelings could still be hurt, even when their bodies could not. They also appeared to have normal empathy for the emotional pain of others.

Although one might imagine that a life without pain would be idyllic, that turns out not to be the case. Pain occurs in response to stimuli that produce tissue damage. Without it, we do not learn to recoil from sharp blades, boiling liquids, or damaging chemicals. People with congenital total insensitivity to pain are constantly injuring themselves. They bite their own tongues, break their own bones, wear down their joints, and scar their corneas by unknowingly rubbing grit into their eyes. Many do not survive beyond their teenage years. Most people with this condition do not die in a dramatic fashion, like the boy in Pakistan who jumped off the roof. Rather, death is often the result of mundane tissue damage: the ill-fitting shoes that damage the feet, the burning food that scars the esophagus, or even the underwear that is too tight and cuts into the skin of the abdomen. Bacterial infection following such injuries is a constant threat.

Brain scans of the affected Pakistani children were normal, as were biopsies of the sural nerve. Unlike the Norrbotten syndrome patients we discussed in chapter 3, these kids had normal numbers of the various sensory fiber types, ranging from the speedy A-alpha fibers to the poky C-fibers. Analysis of their DNA revealed that all six of the children carried mutations in the same gene, SCN9A, which directs production of a voltage-sensitive sodium channel that is essential for the propagation of electrical signals in neurons. However, SCN9A's expression is almost entirely limited to those neurons that convey pain information from the skin and the viscera. (Other neurons utilize different sodium-channel genes.) Consequently, the neurons that convey pain signals to the spinal cord and brain are present, but they are electrically silent. That's why the children's sural nerve biopsies look normal, while their loss of pain sensation is profound. When kidney cells were used to artificially express mutant SCN9A DNA from these patients, no sodium channel function could be seen at all, only a flat line of electrical current (figure 6.1). At present there is no way to restore this current and hence no effective therapy for congenital insensitivity to pain.

The particular mutations in the SCN9A gene that give rise to congenital insensitivity to pain are of a type called loss-of-function mutations: They cause the gene to fail to produce a working protein, resulting in a flat-lined sodium current in pain-sensing neurons. These mutations are inherited in a recessive fashion, so that an individual must have two copies of the SCN9A gene with the relevant mutation, one from each parent, in order to have the disease. This explains why it is much more common in families that practice cousin-marriage. Other types of gene mutation are called gain-of-function mutations. In the case of the SCN9A gene, gain-of-function mutations produce voltage-sensitive sodium channels with aberrant properties. An individual needs to inherit only a single mutant gain-of-function gene to produce a terrible effect on pain perception.

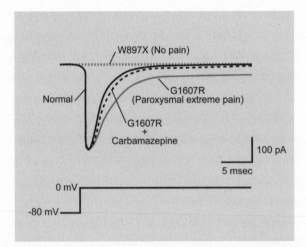

Figure 6.1 Mutations of the SCN9A gene can produce dramatic alterations of pain perception and voltage-sensitive sodium current. Here, different versions of the SCN9A gene have been artificially introduced into kidney cells. Sodium currents are evoked by using an electrical circuit to rapidly shift the voltage across the cell membrane from -80 millivolts to 0 millivolts, roughly mimicking what occurs when a neuron fires an electrical spike to send information along its axon. Normal SCN9A produces a typical sodium current: a brief inward flux of positive current that completely terminates within a few milliseconds. SCN9A from a patient with congenital insensitivity to pain harbors a mutation called W897X, which renders the channel completely nonfunctional, leading to no current at all. SCN9A from a patient with paroxysmal extreme pain disorder bears the mutation G1607R and results in sodium current that activates normally but inactivates slowly and incompletely, leaving a long tail of persistent current. The drug carbamazepine can partially reverse this effect and is useful in reducing the symptoms of paroxysmal extreme pain disorder. There are several different mutations that can give rise to congenital insensitivity to pain, but all produce a flat-line current. Likewise, there are several different mutations that can give rise to paroxysmal extreme pain disorder, but all produce sodium current that inactivates incompletely.

~~~~~~~~

It usually starts soon after birth, often with the first bowel movement: A baby startles, and its face gradually assumes a look of sheer terror. Screaming inconsolably, the newborn clings to an adult. Her body stiffens and flushes strongly, and her face contorts into a grimace. These attacks, which last for minutes and can occur several times a day, are often triggered by some trivial act involving the mouth or the anus: feeding, wiping, or inserting a rectal thermometer. Obviously, babies cannot communicate, but as these children grow, they describe pain that frequently begins in the area of the anus, the jaw, or the eye and then spreads. At once burning, stabbing, and diffuse, the pain is described by adult sufferers as the worst agony imaginable. All the mothers who have this condition reported that it is far worse than going through labor. Most of them said that if they knew their fetus was affected with this condition, they would have aborted it, as they could not bear the prospect of their child enduring such pain attacks.[3]

This condition has been named paroxysmal extreme pain disorder, and it, too, results from a gain-of-function mutation in the SCN9A gene. Figure 6.1 shows sodium current from mutant SCN9A derived from a patient with the disorder. When the cell is depolarized, the sodium current activates normally but inactivates slowly and incompletely. The result is that the pain-sensing neurons are now like machine guns with a hair-trigger: Stimuli that might make normal neurons fire one or two electrical spikes now make them fire a sustained burst. Because of this aberrant electrical signaling, even an innocuous stimulus can trigger a bout of extreme pain. Fortunately, there is a drug called carbamazepine that can promote the inactivation of voltage-sensitive sodium channels, including those derived from the SCN9A gene. In some patients carbamazepine can produce complete relief, and in many others it can reduce the frequency and severity of the pain attacks.[4]

Even without carbamazepine treatment, however, sufferers of paroxysmal extreme pain disorder typically live full lives. Most have children and careers and a normal life span. In a sense this is counterintuitive: If you were forced to choose between two mutant forms of SCN9A, one that would render you painless but would almost ensure that you died young, or another that would give you random attacks of mind-numbing pain for your entire normal life span, which would you prefer?

~~~~~~

Imagine that you're walking around the house without shoes and you slam your toes into a heavy wooden chair leg. The pain doesn't come all at once. First, there's a sharp pain that's localized to the specific toes that you struck, which diminishes quickly. Then you can sing "In-a-gadda-da-vida, baby" before the second wave of throbbing and diffuse pain arrives. The first wave of pain is carried to the spinal cord by a mixture of medium-diameter myelinated A-delta fibers that can transmit electrical spikes at a speed of about 70 miles per hour and large-diameter myelinated A-beta fibers at 150 miles per hour. The second wave of pain is conveyed by small-diameter C-fibers that transmit signals much more slowly, at about 2 miles per hour. All areas of the skin (and most parts of the viscera) are innervated by both fast and slow pain fibers (figure 6.2). The time difference between the first and second waves of pain is most noticeable for locations that are distant from the brain, like the toes. By comparison, facial pain also has fast and slow components, but the gap between when the two are sensed is much smaller, and so the two waves of pain are not as clearly distinguished. Of course this lag time for the second wave of pain is even more pronounced in larger animals. For a ninety-foot-long dinosaur (like a diplodo-cus) who whacked her tail on a floating log, the first wave of pain would have arrived in about one second, while the second wave would have taken a full minute to reach the brain and be perceived.[5]

The initial wave of pain is fast, precise, and discriminative; it provides information related to immediate threat and guides a withdrawal response. Often you've already initiated the withdrawal response and let an expletive fly by the time the second wave of pain arrives. Imagine grabbing a hot pot handle: The first wave of pain causes you to release the handle immediately, and you are already waving your hand in the air to soothe the first pain as the second wave of pain is felt. The second wave of pain is slow to start, slow to end, and is poorly localized. It can have an aching, burning, or throbbing quality. Secondary pain demands sustained attention and motivates behaviors that reduce further injury and promote recovery (like favoring an injured leg while walking).

Pain is not a single, discrete sensation, even when experienced at a single moment in time. We know from our real-life experience that it can have different qualities. In describing the sensory qualities of

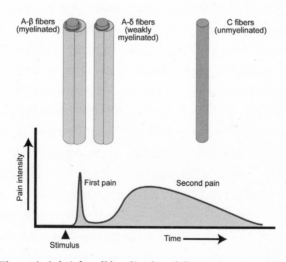

Figure 6.2 First pain is brief, well localized, and discriminative, while second pain is diffuse, emotionally laden, and persistent. First pain is carried by medium-diameter weakly myelin-wrapped A-delta fibers and large-diameter myelin-wrapped A-beta fibers, while second pain is conveyed by unmyelinated small-diameter C-fibers. One of the ways we know this is by using a ligature that will compress and block the A-fibers while sparing the C-fibers. This will eliminate the experience of first but not second pain. In case you're wondering, the SCN9A gene is expressed in both the A-fiber and C-fiber pain neurons, so people with congenital total insensitivity to pain lack both first and second pain and people with paroxysmal extreme pain disorder have enhanced first and second pain.

pain, people use terms like *sharp, throbbing, flashing, burning, tingling, dull, aching, heavy,* and *stinging* (figure 6.3). How do these different types of sensation emerge? There are three main categories of pain sensor: mechanical, thermal, and polymodal. We conjecture, but don't really know, that the particular sensory qualities of pain arise from both the pattern of neural activity and the relative degree of activation of each of these three pain sensors, perhaps combined with or compared to nonpain touch signals from the same location on the body.

Unlike the sensors for pressure, vibration, texture, and caress that have specialized structures or elaborate associations with hair follicles, pain-sensing neurons use simple, unadorned free nerve endings. In the skin these free nerve endings penetrate the epidermis. Mechanical pain sensors are most easily activated by intense pressure: If you cut your finger with a knife, or stub your toe while walking, or pinch your skin in a zipper, your mechanical pain sensors will send signals to your brain. Some of these sensor molecules are embedded within the free endings of A-delta fibers, so this information will be conveyed quickly. Thermal pain sensors are also located in the free endings of a different group of A-delta fibers and respond to temperatures below approximately 42°F or above approximately 115°F.[6] There is also a subset of C-fiber endings that respond more broadly—to thermal, mechanical, or chemical stimulation (like strong acids or bases). These polymodal pain sensors are responsible for the second wave of pain, and their wider susceptibility to different types of pain helps to explain why this second wave is less qualitatively specific than the first.[7]

The cell bodies of the A- and C-fibers that carry pain signals are located in the dorsal root ganglia and enter the spinal cord in a region called the dorsal horn (figure 6.4). This is similar to the anatomy of the sensory nerves for fine touch and caress, as discussed in chapters 2 and 3. The C- and A-delta pain fibers make excitatory connections with neurons located in the dorsal horn. There are

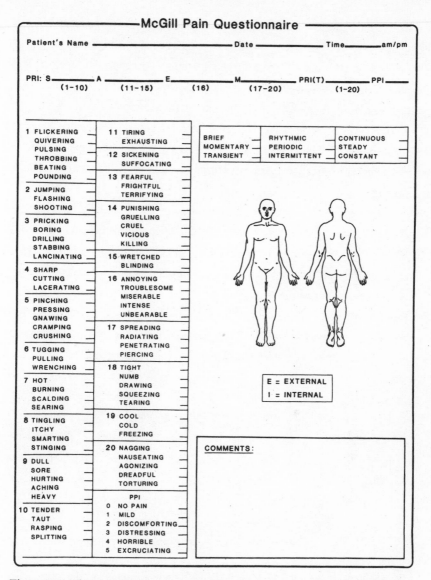

Figure 6.3 The McGill Pain Questionnaire was developed by Dr. Ronald Melzack to try to encompass the variety of pain experience in a clinical setting.[8]
Groups 1–10 are sensory descriptors, 11–15 are emotional, and 16 is evaluative.
Groups 17–20 are miscellaneous and have aspects of all three of the other categories. Copyright R. Melzack, 1975; reprinted with permission.

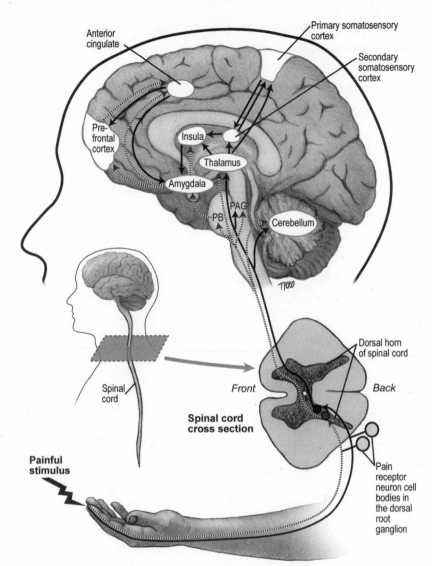

Figure 6.4 Two major pathways conveying pain information to the brain. The fast sensory-discriminative pain pathway (solid black line), mostly carried in the spinothalamic tract, passes through the thalamus to engage the primary and higher somatosensory cortices. The slow affective-emotional pathway (dashed line) runs in part through the spinomesencephalic tract, via the parabrachial nucleus, and engages the amygdala, the insula, and the anterior cingulate cortex.
© 2013 Joan M. K. Tycko

different layers within the dorsal horn, containing neurons that receive different types of touch information, including proprioception (carried by superfast A-alpha fibers), fine mechanosensation
(carried by fast A-beta fibers), caress (carried by a different population of slow C-fibers), and pain. For our purposes it's not necessary
to examine the detailed anatomy of these layers.[9] There is an important general principle to keep in mind, however: While the various streams of different types of touch stimuli (pain, fine touch,
caress, etc.) are generally kept separate while coursing through the
spinal cord to the brain, there is some notable mixing of signals.[10]
For example, a type of neuron in the spinal dorsal horn called a
wide-dynamic range neuron integrates pain and fine touch information. This blending of pain with nonpain signals in the spinal
cord may account for why an action like rubbing a smashed elbow
can temporarily dull the pain of that injury.

These wide-dynamic range neurons' integration of various types
of pain signal can at times lead to sensory illusions. For example,
some wide-dynamic range neurons in the spinal dorsal horn receive
pain information from both the viscera and the skin. Sufferers of
angina (pain from insufficient blood flow to the heart muscle) often
experience pain that feels as if it were coming from the left arm,
even though that limb is uninjured. This "referred pain" is a clear
demonstration of the general principle that we cannot always accurately decode the sensory world. In this case, the wiring diagram of
the spinal cord is built in a way that creates confusion—which begs
the question: Is there an advantage to having pain signals converge
and create ambiguity? The short answer: We don't know.

~~~~~

There is no single brain area that is responsible for registering pain.
Rather, pain perception is distributed over a group of brain regions, each involved in a different aspect of the pain experience
(figure 6.4). There are at least five different neural pathways that
carry pain information from the neurons of the spinal dorsal horn,

but we'll focus on only three of these.[11] The first, the spinohypo-thalamic pathway, activates the hypothalamus, a structure at the base of the brain, to produce rapid, subconscious, pain-evoked alterations in heart rate, body temperature, breathing, core muscle contraction, and hormone secretion. The fibers of the second, the spinothalamic pathway, originate in neurons in the spinal dorsal horn, cross the midline, ascend in the spinal cord, and then make synapses in the thalamus. The thalamic neurons in turn send fibers to the primary and secondary somatosensory cortices. If an electrode is placed in the spinothalamic tract and a few of its fibers are briefly activated, a highly localized, well-defined painful sensation will be evoked.

When brain scanning was performed in association with a painful stimulus, it was found that the first wave of pain was primarily correlated with activation of the spinothalamic tract and its targets, the primary and secondary somatosensory cortices. The second wave of pain was most clearly associated with activation of a third ascending pathway, called the spinomesencephalic tract, which activates the parabrachial nucleus in the brain stem and, through further synaptic relays, the insula, amygdala, and anterior cingulate cortex.[12] Why are such neuroanatomical details important? The reason is that these spinomesencephalic tract targets in the brain are involved in emotional and cognitive pain responses. Their activation does not encode the precise location or quality of pain, but, rather, gives pain its characteristic negative emotional tone. They also integrate pain sensation with other information about the situation at hand: Am I safe or under threat? Was that pain anticipated or a surprise? What are the future implications of this pain?[13]

We experience pain as intrinsically unpleasant. In describing pain, people use emotional words like *punishing*, *cruel*, and *unbearable* (see the McGill Pain Questionnaire, figure 6.3). When pain occurs, we don't experience it as a series of distinct sensory-discriminative and emotional-affective components: We experience it as a unified, unpleasant sensation. The emotional and the sensory are completely blended. However, when certain types of brain damage occur, the component parts of pain can be distinguished by affected individuals.

Selective damage to the lateral portion of the thalamus and the primary and secondary somatosensory cortices results in a syndrome in which the sensory-discriminative properties of the pain stimulus are lost. Incredibly, people with this type of damage can describe an unpleasant emotional reaction to a painful stimulus but are completely unable to resolve the quality of the pain (burning versus freezing, sharp versus dull) or even specify its location on the body. Conversely, selective damage to the posterior insula or the dorsal anterior cingulate cortex, central nodes of the affective-emotional pain circuit, can result in a condition called pain asymbolia. These patients are able to accurately report the quality, intensity, and bodily location of the painful stimulus, but they lack the negative emotional response to pain that the rest of us take for granted. Because pain asymbolics no longer appreciate the destructive significance of pain, they are slow to withdraw from a painful stimulus. They feel pain, but it just doesn't seem to bother them:

> Pricked on the right palm, the patient smiles joyfully, winces a little, and then says, "Oh, pain, that hurts." . . . The patient's expression is one of complacency. The same reaction is displayed when she is pricked in the face and stomach. Pricked on the soles of the feet, she begins to smile, openly titillated.[14]

Pain asymbolics are not masochists; in fact, they are the opposite of masochists, for whom pain has deep emotional meaning. Pain asymbolics do not enjoy pain and do not seek it out. Nor are they merely spacey and inattentive. Pain simply has no emotional resonance for them, either positive or negative.

Pain is intrinsically emotional and negative, in much the same way that orgasms are intrinsically emotional and positive. Both the normal experience of orgasm and the normal experience of pain require the near-simultaneous activation of several brain regions to produce the sensation that we experience as a unified whole. Pain and orgasm require the primary and higher somatosensory cortices for the sensory-discriminative portion and another region for the affective-emotional portion: the posterior insula, anterior cingulate cortex, and associated regions for pain; and the ventral tegmental area and the targets of its dopamine neurons for pleasure. Stripped of their emotional components, pain and orgasm are rather tepid, reflexive experiences.

~~~~~~

When I was a child, sunburn seemed like sympathetic magic. I'd spend the day at the beach, roasting in the sunshine and splashing in the waves. Then, that evening, the heat of the sun's rays would follow me indoors, trapped in my skin, keeping me up at night and making the light touch of the bedsheets and the spray of a hot shower unbearable. Sunburn produces allodynia, a painful sensation in response to touch stimuli that are normally innocuous— like the light stroking of sunburned skin. Allodynia shares many features with another form of persistent pain called spontaneous pain, which occurs when pain is felt in the absence of any specific stimulus to the body. There are two key features of persistent allodynic pain and spontaneous pain arising from tissue damage. The first is that these forms of persistent pain will become generalized. Tissue damage from burning, for example, will make the damaged area more sensitive not only to heat but also to mechanical stimulation: If you burn the distal pad of your thumb while cooking and then try to grasp a pen to write, the innocuous mechanical stimulation will also cause pain. The second is that the inflammation that occurs in response to tissue damage (including the symptoms of swelling, redness, and the sensation of heat) will not be precisely restricted to the damaged tissue but will spread for some distance beyond it. A minor burn to the pad of your thumb, for example, may well cause the entire digit to become inflamed for days; this inflamed area, and even a bit of tissue beyond that area, will experience allodynia and spontaneous pain.

Inflammation and the persistent pain associated with it are produced by a complex brew of chemical signals called the inflammatory soup (figure 6.5). When tissue is injured, its damaged cells release a set of compounds called prostanoids, which can affect receptors like TRPV1 on the endings of C-type pain fibers. Damaged tissue can also activate white blood cells, like mast cells and macrophages, causing them to release a compound called bradykinin, which, like prostanoids, reduces the temperature threshold of TRPV1 activation from a toasty 109°F to an otherwise harmless 85°F (as discussed in chapter 5). Other compounds released from macrophages, like the proteins TNF-alpha and NGF, also act to sensitize C-type pain fibers. Activated mast cells release histamine, which targets blood vessels to dilate them and make them slightly porous to blood plasma, thus resulting in warmth, redness, and swelling of the surrounding tissue.

Initially it was thought that the nerve fibers were merely the recipients of these painful chemical signals. Now it is well established that the terminals of the pain-sensing C-fibers also signal back to the tissue in a positive feedback loop. Nerve terminals release a molecule called CGRP, which promotes blood vessel dilation and plasma leakage. They also release another molecule called substance P, which activates mast cells. The ongoing flow of chemical signals among damaged tissue, white blood cells, blood vessels, and pain-sensing C-fibers is one of the reasons why pain and inflammation can persist for days to weeks following an injury. Because these chemical signals can diffuse to neighboring healthy tissue and trigger the feedback signaling there, swelling and hyperalgesia can spread, but only in a limited fashion: While damage to your thumb may cause your hand to swell and ache, in the absence of infection, it probably won't cause your whole arm to do so (figure 6.5).

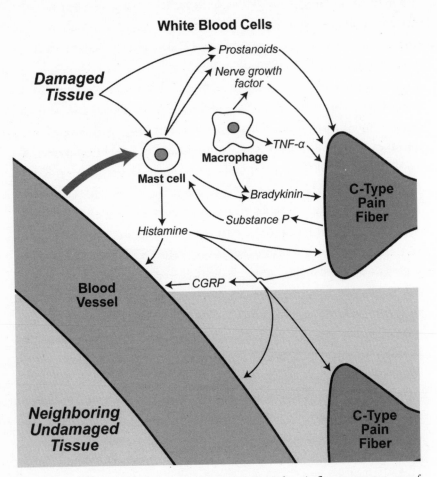

Figure 6.5 Tissue damage results in production of an inflammatory soup of chemical signals. Some signals derive from the damaged tissue itself (like keratinocyte cells in skin), while others originate from white blood cells (macrophages and mast cells), and yet others are secreted by the C-type pain-sensing nerve fibers. The end result of all this is a positive feedback loop that causes pain and inflammation to persist and spread, in part through the diffusion of the signaling molecule histamine. Drugs to blunt pain and inflammation often act to block certain parts of this chemical-signaling network. While this diagram may look complex, and it is, there are many additional components of the inflammatory soup that are not shown here.

Many of our most useful drugs for treating pain and inflammation act on chemical signals in the inflammatory soup. Aspirin, acetaminophen (Tylenol), and ibuprofen all inhibit the production of prostanoids. Antihistamines block the action of histamine on its receptors on nerve terminals and blood vessels. In recent years drugs that interfere with TNF-alpha signaling have revolutionized the treatment of rheumatoid arthritis pain. Drugs that disrupt the action of NGF hold great promise of relieving persistent pain and are presently in clinical trials, but they seem to accelerate the degeneration of the joints, so it is unclear whether they will ultimately be effective.[15] Compounds extracted from pineapple and aloe can interfere with the action of bradykinin. It's possible that they may have therapeutic uses or may serve as a template for the design of bradykinin-blocking drugs inspired by the structure of these natural compounds. Developing drugs to interfere with additional compounds in the inflammatory soup will continue to be an important endeavor.

~~~~~

Francis McGlone, a well-known researcher on tactile perception, is fond of asking, "Why is it that there is chronic pain but no chronic pleasure?" It's a good question. We've seen how pain is essential for inciting behaviors that will minimize tissue damage, and how those who lack pain perception rarely survive past childhood. Yet pain can also outlast its usefulness, tormenting people long after tissue damage has healed, sometimes for a lifetime.

Persistent pain is produced not only by changes to the endings of pain-sensing fibers. There are also alterations at the synapses in the spinal cord, where these fibers contact the neurons of the spinal dorsal horn. When electrical spikes are conveyed to the central terminal of pain-sensing C-fibers, they cause release of the excitatory neurotransmitter glutamate. Glutamate diffuses across the tiny synaptic cleft separating the two neurons and binds to glutamate receptors on the dorsal horn neuron, resulting in the propagation of pain signals through the spinal cord and on to the brain. When this synapse is repeatedly stimulated, as in the case of persistent pain, it becomes stronger and more efficient. This results from a number of factors, including increased glutamate release, more glutamate receptors on the dorsal horn neuron, as well as changes to the voltage-gated ion channels in the dorsal horn neuron that promote repeated spike generation.[16] These changes can be very long lasting, like memories. (Indeed, it is thought that some of the molecular and cellular changes that encode memories in the brain are similar to those that underlie this form of chronic pain originating in the spinal cord.) Even when the inflammatory response in the skin (or other tissue) has completely faded and the pain-sensing capacities of the nerve endings have returned to normal, changes in the spinal cord can endure for months or years. In some cases, chronic pain originating in the spinal cord can last a lifetime. It would be enormously beneficial to find a way to selectively weaken these pain-transmitting synapses in the spinal dorsal horn, and, not surprisingly, it is an active area of research.

One type of persistent pain that is particularly resistant to treatment is phantom limb pain. Following amputation, about 60 percent of patients experience a feeling of chronic pain in the limb that was removed. Sometimes it's an aching; at other times it's experienced as burning. The phenomenon is more likely to occur when amputation occurs in adulthood, and is equally likely to result from a surgical amputation as it is from a traumatic loss of limb. Originally it was thought that phantom limb pain resulted from damaged nerve endings in the stump, but neither repeated surgery on the stump nor local anesthetics delivered to the area have been shown to relieve it.

It is likely that at least a portion of phantom limb pain is a function of persistent strengthening of pain fiber-dorsal horn neuron excitatory synapses. For years amputation surgery was performed solely under general anesthesia. In this scenario, the pain signals from the procedure travel from the periphery to the spinal dorsal horn but are blocked at later stages from reaching the brain. The incidence of phantom limb pain appears to be somewhat reduced when, in addition to a general anesthetic, local anesthetics are given to numb the amputated region prior to and during the amputation surgery, preventing the pain signals from reaching the spinal dorsal horn at all. Drugs that inhibit the persistent strengthening of the synapses received by the dorsal horn neurons have also shown some limited promise for reducing the incidence of phantom limb pain when used in this fashion.[17]

When the synapses that convey pain information in the spinal cord are persistently strengthened, the signals that spinal cord neurons send to the brain will be altered as well, as, in turn, will the brain itself. In one sense the situation is much like that we discussed in chapter 2, where the representation in the body map of the fingering hand of experienced string players was expanded following years of practice. Patients with amputations who experience phantom limb pain sometimes show enlarged representation of both the affected limb area and other regions in the body map of the primary somatosensory cortex. Amputees who do not suffer from chronic pain, however, do not show these enlargements.[18]

~~~~~~

On April 13, 2003, Private Dwayne Turner, a U.S. Army combat medic, was with a small unit that came under attack as they were unloading supplies in a makeshift operations center about thirty miles south of Baghdad. Recounting that day in an interview many months later, Turner recalled:

> . . . a[n enemy] guy lugged a grenade over one of the walls, and me and some other guys were right in the middle of the blast. I ran to the front of the vehicle and I saw some of the wounded, and I kind of sat back and assessed the situation because there were still rounds going around. I was like, well, there's guys going down and,

you know, I'm not just going to leave them there. I got to
doing my thing, so I started going to work...I know
these guys. I know them, know them back home, we eat
with these guys, we sleep with these guys, and you know,
you just don't want nobody to die on your watch. They're
pretty much your brothers. You're going to war with your
family. I know that if my real life brother was out there,
I'd go and save him.

Turner had taken shrapnel from the grenade blast in his right leg,
thigh, and abdomen, but it didn't slow him down very much. He
dashed out from cover repeatedly to retrieve his fallen comrades and
pull them to safety, and in the course of doing so was shot twice, one
bullet hitting his left leg and another breaking a bone in his right
arm. He barely noticed that he'd been shot.

I saw bullets going around and I thought actually that
the bullets were hitting the ground and they were, you
know, dust or rocks were popping up and hitting me, because I felt like little nicks in certain places. But other than
that, I really didn't know until somebody told me, he was
like, "Doc, Doc, you're bleeding." I was like, no, that's
not me, that's somebody else. Trying to pass the buck, but
I guess it was [me].[19]

Private Turner eventually collapsed from blood loss, but a few
minutes later he had to be restrained from returning to help his fellow soldiers. He and the other wounded soldiers were later evacuated by helicopter. The army, in awarding Dwayne Turner the
Silver Star for valor, estimated that at least twelve soldiers would
have died without his actions.

While it is remarkable that in the heat of combat Private Turner
was able to ignore the pain from his shrapnel wounds and did not
even realize he'd been shot, it does not diminish his heroism at all

to note that ignoring severe pain in the heat of battle is not un-usual. Lieutenant Colonel Henry Beecher compiled statistics on pain in the wounded soldiers he treated during the Allied invasions of Italy and France in World War II. He wrote that about 75 per-cent of badly wounded men reported such little pain that they re-fused pain-relief medication when offered it on the field of battle. Yet a few days later, when recovering in a hospital, these same pa-tients protested vigorously at the minor pain of an inept blood draw, just like anyone else would.[20] Soldiers are not superhumans with absurdly high pain tolerances. Rather, they are ordinary people who are thrust into extraordinary situations in which the cognitive and emotional stressors of combat serve to blunt pain.

When I was five years old, going to the pediatrician to get a rou-tine immunization seemed like a kind of death sentence. Like Pav-lov's dog's, my heart would speed up as soon as my mother's car neared the doctor's office. The blond wood paneling of the hall-ways in the medical building, the tap of my mother's heels on the tile floor, and the smell of alcohol pads all reminded me of the last horrible injection. When the time for the procedure arrived, I was so laser-focused on that one patch of skin on my left arm that my eyes bugged out. The tiny prick of the needle felt as if a steak knife had been thrust deep into my muscle and twisted. As I howled in pain, the doctor observed drily, "What a dramatic kid—a real Sarah Bernhardt."

As a child, I experienced many minor injuries that were equally or even more painful, from skinned knees to a bumped head. But those came as surprises; because they weren't anticipated, they didn't seem such a major concern. The brief and minor pain of the injection, however, was amplified enormously by my mounting fearful expectations and my memories of the previous traumatic injections.

These accounts of heroic combat and pediatric melodrama dem-onstrate how cognitive and emotional factors can dull or heighten pain perception. Can we understand this cognitive and emotional modulation of pain perception in terms of the distributed network of pain-processing centers in the brain? In general, the answer is yes, but many of the details remain to be worked out. One key insight is that the brain can send signals down to the pain-transmitting neurons in the dorsal horn of the spinal cord that can say either, "Speak up and say it louder!" or, "Shut up! Dial down the pain information!" The truly amazing fact is that *the brain is exerting control over the information that it receives*. It is not just taking in all the data and then biasing its perceptions and responses based upon the present emotional or cognitive state; rather, through these descending nerve fibers, it is *controlling* which sensory information will be received from the spinal cord. This is a weird and counterintuitive state of affairs. The brain actively and subconsciously suppresses or enhances pain information on a moment-to-moment basis. It spins the media, so to speak. This realization that in many cases we have access only to self-censored information is somewhat disconcerting to those of us who like to feel that we have access to unfiltered reality to guide our rational thoughts.[21]

As we have seen, information from both the sensory-discriminative and the affective-emotional regions of the brain's pain-processing network converges in several locations, including the anterior cingulate cortex, the insula, the prefrontal cortex, and the amygdala. These regions then send signals to a structure in the brain stem called the periaqueductal gray (figure 6.6), which in turn excites structures farther down in the brain stem called the locus coeruleus and the rostroventral medulla.[22] It's these latter regions that finally send axons down to the dorsal horn of the spinal cord, where they form synapses that can either suppress or boost signaling from peripheral pain-sensing fibers. There are some cells in the rostroventral medulla, called on-cells, that increase firing, and others, called off-cells, that decrease their firing in response to pain. Increases in on-cell activity boost transmission of pain signals in the spinal dorsal horn and enhance pain perception, while increases in

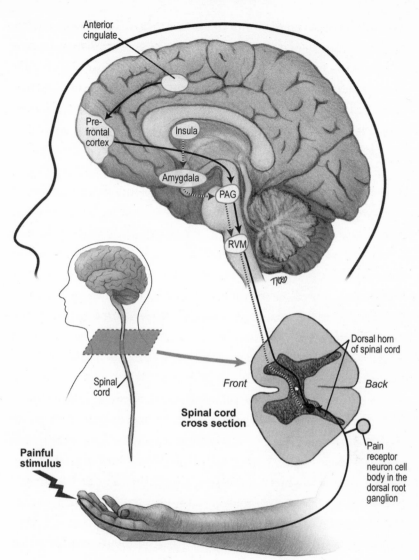

Figure 6.6 Descending pathways, running from the brain to the spinal cord, have an important role in the cognitive and emotional modulation of pain. The key way stations for this descending information in the brain stem are the periaqueductal gray region (PAG) and the rostroventral medulla (RVM). This diagram is a simplified representation. In particular, the synaptic actions of several neurotransmitters released in the spinal dorsal horn by descending fibers are varied. © 2013 Joan M. K. Tycko

off-cell activity have the opposite effect.[23] It is this circuit that enables the brain to dial the incoming pain information up or down.

The pain-controlling properties of the opium poppy and its derivatives, like morphine, have been known since at least 3400 BC in Sumeria (present-day southern Iraq). The Sumerians called the poppy Hul Gil ("joy plant") and soon exported it to Assyria and Egypt, from which it spread widely. Morphine works by mimicking the action of the brain's own morphinelike molecules, the endorphins and enkephalins. The receptors for these opioids are widespread in the body and in the nervous system, but a particular type, called the mu-opioid receptor, is concentrated in the descending pain pathways, in the periaqueductal gray, the rostroventral medulla, and the superficial layers of the spinal dorsal horn.[24] Microinjecting morphine into the periaqueductal gray region is sufficient to produce a strong analgesia, mediated by exciting off-cells and inhibiting on-cells in the rostroventral medulla. The pain-blunting effect of the periaqueductal gray region is so strong that carefully controlled electrical stimulation of this region can replace chemical anesthetic agents during surgery. Some people suffering from pain that does not respond well to drugs have had electrodes implanted in their periaqueductal gray area to enable them to self-stimulate their brains with a handheld device for pain relief. Pain relief via periaqueductal gray self-stimulation can be blocked by drugs that interfere with mu-opioid receptors, implicating endorphins/enkephalins in the analgesic effect.

Armed with this knowledge of descending pain modulation, let's return to our examples of Private Turner, who, in the heat of battle, didn't realize that he had been shot, and your youthful narrator, crying out at the minor pain of a routine injection. Clearly, the behavior in both cases was in part a matter of attentional focus. Private Turner was focused on saving his buddies while under fire; his thoughts were not on his own body. In the laboratory, when subjects are distracted from a pain stimulus (such as by being asked to respond to questions or perform a mental task), their rating of the pain intensity decreases. This response is associated with decreases in activity in the primary somatosensory cortex and the insula. Conversely, when subjects are instructed to focus on their pain, their ratings of pain intensity rise, reflecting increased activity in these same brain regions. Importantly, the emotional component of pain perception, indicated by ratings of pain unpleasantness, is largely unaltered by focused attention.[25]

In the confines of the laboratory, we can design experiments that selectively engage the attentional and emotional mechanisms. However, in the real world, these are not so easily separated. Private Turner was distracted by the ongoing firefight and his efforts to save his fellow soldiers, but this diversion was not emotionally neutral. Rather, it was fraught with emotions: fear, compassion, pride, and his deep bonds with his fallen comrades. In the doctor's office, as I waited for the injection, my focus on the impending prick of the needle was not emotionally neutral either, but laced with dread, seeded by frightening memories.

Negative emotions enhance perception of pain, but they do so in a way that is perceptually and anatomically different from attentional effects. Negative emotions boost pain-evoked activity in the anterior cingulate cortex and increase the ratings of pain unpleasantness, but not their intensity. The end result is that both the sensory-discriminative and the affective-emotional aspects of pain perception are subject to cognitive and emotional modulation. And in the real world, the cognitive and the emotional are highly interdependent. For example, pain unpleasantness, when endured persistently, engages the prefrontal cortex, a region involved in rumination and reflection about the future implications of persistent pain: How long will this pain endure? Will the pain start again? Is it under my control or not? How safe am I? These prefrontal cortex-based ruminative processes contributed to my youthful anxiety about the impending injection and exerted their effects, in part, through descending pathways to the spinal cord. In many people, this sets up a vicious cycle, a positive feedback loop in which rumination boosts pain unpleasantness, which in turn triggers additional anxiety and rumination. This is why tranquilizers (like the benzodiazepines) can be useful in treating pain, particularly chronic pain. Even if these drugs don't affect pain perception directly, they reduce anxiety and therefore help to attenuate pain unpleasantness and break the positive feedback loop. On average, people suffering from mood disorders are at significantly greater risk for developing chronic pain.[26]

When pain activates the spino-hypothalamic pathway, it produces a group of fight-or-flight responses, including increases in heart rate, sweating, breathing rate, core muscular contraction, etc. We can sense these responses both consciously and subconsciously, and they contribute to a sense of anxiety. When you are aware of your heart beating faster, you feel more uneasy and threatened. This is another positive feedback loop that can contribute to chronic pain: Subconscious fight-or-flight responses evoked by pain boost anxiety, which in turn increases pain.

The portion of the brain's pain circuit including the insula, the anterior cingulate cortex, and the prefrontal cortex is crucially involved in comparing one's expectations for pain with pain sensations as they are actually happening, and thereby generating predictions about the future. One key factor that is evaluated is threat, which originates externally. Pain that is self-generated is less threatening and is rated as both less intense and less unpleasant than pain that intrudes unexpectedly from the outside world. In the laboratory, self-generated pain is associated with reduced activation in both the primary somatosensory cortex and the anterior cingulate cortex when compared with the same pain stimulus generated externally. In one study this was demonstrated by having either the subjects themselves or the experimenter squeeze one hand with the other to produce compression force against pointed plastic beads.[27]

In other words, while it might hurt to accidentally hit your thumb with a hammer, at least you know how it happened, and you feel as if you can control the situation to reduce the chance of it happening in the future. That knowledge reduces the sense of threat and thereby attenuates the pain. Conversely, not knowing when a pain stimulus will occur again boosts both the intensity and the un-pleasantness of the experience.

~~~~~~

When patients are given a sugar pill or some other placebo and told that it will relieve pain, many will indeed experience a degree of pain relief (figure 6.7). While the prevalence and degree of the relief vary with the type of pain and the personality of the patient, on average, the placebo effect evokes significant analgesia in about 30 percent of patients. The underlying mechanism of the placebo effect is complex and probably involves both anxiety-reducing and unpleasantness-reducing components. Brain-imaging studies have shown that placebo analgesia is associated with the release of en-dorphins/enkephalins in several parts of the emotional pain circuit, including the anterior cingulate, prefrontal cortex, and amygdala, as well as the periaqueductal gray region. Naltrexone, a drug that blocks mu-opioid receptors, also blocks placebo analgesia, sug-gesting that the action of endorphins/enkephalins is required.[28] A variant of the placebo effect can also be engaged to *increase* per-ceived pain: If patients are told that they are receiving an analgesic treatment that is unlikely to work, then fewer people will report pain relief, even if they are administered very effective analgesics like oxycodone or morphine. Or if people are given an inert con-trol treatment but are told that it will increase pain, then, on aver-age, subjects will report a heightened pain sensation. This is called the nocebo effect, and its biological basis is poorly understood, but it appears to be associated with increased activation of pain percep-tion regions in both the spinal cord and the brain.[29]

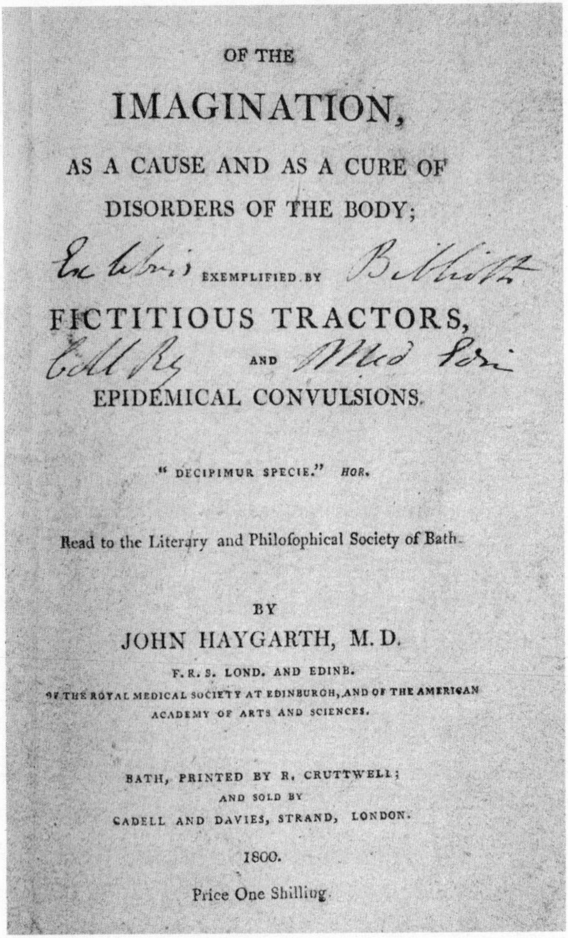

OF THE

IMAGINATION,

AS A CAUSE AND AS A CURE OF

DISORDERS OF THE BODY;

EXEMPLIFIED BY

FICTITIOUS TRACTORS,

AND

EPIDEMICAL CONVULSIONS.

" DECIPIMUR SPECIE." HOR.

Read to the Literary and Philosophical Society of Bath.

BY

JOHN HAYGARTH, M.D.

F. R. S. LOND. AND EDINB.

OF THE ROYAL MEDICAL SOCIETY AT EDINBURGH, AND OF THE AMERICAN
ACADEMY OF ARTS AND SCIENCES.

BATH, PRINTED BY R. CRUTTWELL;
AND SOLD BY
CADELL AND DAVIES, STRAND, LONDON.

1800.

Price One Shilling.

Figure 6.7 John Haygarth was the first person to systematically investigate and report on the placebo effect. His book, published in 1800, examined the efficacy of "Perkins tractors," a type of pointer touted as having the ability to draw out disease through their special metal tips. Haygarth showed that some people did improve after treatment with Perkins tractors, but they improved to the same degree when treated with fake Perkins tractors made of wood. His conclusion: Perkins tractors are a sham, but "the passions of the mind had a wonderful and powerful influence upon the state and disorder of the body." If you think you'll get better, then you are more likely to do so. The book's epigraph, *"Decipimur specie,"* is an abbreviation of *"Decipimur specie recti,"* a well-known quotation by the Roman poet Horace, which translates as "We are deceived by the appearance of rectitude." Used with permission of the Royal College of Physicians, Edinburgh.

Torturers are all too aware of the cognitive and emotional modulation of pain perception, which they exploit in horrifically dehumanizing ways to heighten pain and fear and leave their victims feeling powerless. Because they know that anticipation of pain can increase the experience of it enormously, they will often let their victims see and hear others being tortured to engage that anticipation and focus. Torturers likewise understand that the experience of pain is heightened by a sense of threat. This is accomplished by subjecting their victims to humiliation (stripping, sexual abuse, screaming) and to an irregular and unpredictable schedule of sleep/waking/feeding and torture.

Fortunately the same emotional and cognitive control circuitry that is exploited by torturers to maximize pain can potentially be used beneficially to reduce pain, particularly chronic pain. A number of mindfulness-based practices—including meditation, yoga, Tai Chi, and the Feldenkrais Method—have been reported to decrease both chronic and acute pain sensations. Although the quality of some of these studies has been low, lacking appropriate controls and enrolling too few subjects to allow for strong statistical analysis, there are indications from some large, randomized, well-conducted studies that these techniques can be efficacious for certain forms of chronic pain.[30]

How might mindfulness-based training serve to attenuate pain? Let's examine this in the context of meditation, for which there is useful literature examining pain perception and its neural processes. One general way to reduce pain would be to learn to exert some control over subconscious pain-evoked fight-or-flight responses, thereby attenuating pain-evoked anxiety and threat and reducing pain unpleasantness.[31] This would interrupt one of the positive feedback loops that sustain chronic pain. Another way would be to learn to experientially open oneself to pain. Rather than avoid or reject the pain experience, one would learn over time that repeated pain is not a threat, and this nonreactive, nonjudgmental learning would then result in decreased pain unpleasantness. Indeed, recent studies from the group of Richard Davidson at the University of Wisconsin, Madison, have shown that expert Buddhist meditators (more than ten thousand hours of practice) engaging in open presence meditation report equal pain intensity but reduced pain unpleasantness in response to a heating laser when compared with novices. When brain scanning was performed, it revealed that expert meditators had reduced baseline activity but increased initial pain-evoked activity in the insula and the anterior cingulate cortex, which was attenuated with repeated stimulation. The authors suggest that meditative training to develop experiential openness decreases the anticipation of pain and increases the recruitment of useful attentional resources during pain to allow for effective learning. This learning would reduce the perception of threat, the related anxiety, and the concomitant fight-or-flight responses.[32]

Zen is a different form of meditation. Rather than opening oneself to experience, one seeks a more detached form of self-regulation. The goal is not learning but to dissociate oneself by reducing higher-order evaluative processes. When Pierre Rainville's group at the University of Montreal studied experienced Zen meditators with a thermal pain stimulus, they found a result that was similar to that of the Davidson group: Experienced meditators reported reduced pain unpleasantness and increased pain-evoked activity in the insula and anterior cingulate cortex as compared with nonmeditators. Crucially, they also showed reduced activity in the prefrontal cortex, an effect that was strongest in those Zen meditators who had the lowest pain unpleasantness scores. One interpretation of these imaging results is that the Zen meditators are not learning to conceptualize repeated pain as a nonthreat (which would presumably require the involvement of the prefrontal cortex) so much as they are simply deciding to ignore it.[33] It is likely that there are multiple cognitive/emotional strategies that can serve to reduce pain.

~~~~~~

Pain and negative emotion are deeply intertwined. In everyday usage we speak of social rejection by peers, family, or even strangers as "hurtful" and rejection in love as "heartbreak." You'll recall that this is not our first foray into tactile metaphors. We've examined how the sense of touch is intrinsically emotional (feelings!) and how social warmth and physical warmth are interrelated. In the cases of cool mint and hot chili peppers, we've even seen how the metaphor is encoded in the sensor molecules TRPM8 and TRPV1.

We know that pain perception has an anatomically distinct emotional component that can be disrupted by damage to the insula or the anterior cingulate cortex. But what exactly *is* the relationship between physical and emotional pain? There are some suggestive correlations. Studies have shown that those individuals whose feelings are easily hurt, particularly from social rejection, also tend to rate physical pain as more unpleasant when tested in the laboratory. And even in people whose feelings are not easily hurt, experiences that heighten social distress can increase the perceived unpleasantness of physical pain. Surprisingly, analgesic drugs—even fairly mild ones like acetaminophen (Tylenol)—can reduce social pain. But perhaps the most compelling finding is that social rejection and physical pain activate overlapping regions of the brain's emotional pain circuit: When subjects are excluded from participation in a virtual ball-throwing task on a computer, even this mild form of social ostracism produces activation of the dorsal anterior cingulate cortex and the anterior insula. Another study analyzed a much more potent form of social rejection: When people who had recently been dumped were asked to look at a photo of their former beloved, not only were emotional pain centers engaged but, in addition, the secondary somatosensory cortex, a sensory-discriminative pain center, was also activated.[34]

Once again, everyday language is reflective of neural processes. The similarity of emotional pain and physical pain is not merely a construction of evocative or poetic speech. The metaphor is real and it is encoded in the brain's emotional pain circuitry. Social rejection hurts.

CHAPTER SEVEN

THE ITCHY AND SCRATCHY SHOW

Semanza lived in the Rukungiri district of rural Uganda. He suffered such unbearable itching that continuous scratching with his fingernails did not afford him even temporary relief. His solution was to break a clay pot and use the rough edge of one of its pieces as a scratching tool. Eventually his skin became severely damaged and infected with bacteria. Years of relentless itching and scratching had left it so calloused that syringe needles could not penetrate it. Moses Katabarwa, an epidemiologist and health worker at the Carter Center's River Blindness Program who met Semanza in 1992, said that his skin appeared to be covered in dried mud. No one from his village wanted to be near him, and so Semanza, shunned, lived in a small hut behind his family's home.

The source of Semanza's unbearable itch was onchocerciasis, infection with a parasitic roundworm called *Onchocerca volvulus*. Because this infection can sometimes target the eye and the optic nerve, it is also known as river blindness. This worm is transmitted in larval form by the bite of a black fly that thrives amid fast-moving tropical streams. The disease is not directly produced by the worm but rather by a bacterium that dwells in the gut of the worm and is released when the worm dies, triggering an immune reaction by the human host.

About 18 million people have contracted onchocerciasis, almost all of whom live in Africa, with a few scattered cases in Venezuela and Brazil. Onchocerciasis is not fatal but it results in a miserable life. The disease has blinded about 270,000 people alive today. In Liberia, infected workers on a rubber plantation have been known to place their machetes in a fire pit and then use the red-hot blades as a tool to relieve the relentless itching. Of course the itching also makes sleep elusive, and as Moses Katabarwa explains, "Children with the worms can't concentrate because they are scratching themselves all day and night." Suicide is common among its victims.

While there is no vaccine for onchocerciasis, it can be controlled with a drug called ivermectin, which has been donated worldwide by the pharmaceutical firm Merck since 1985. Treatments with ivermectin every six months kill newborn worms (called microfilariae), which releases the itch-triggering bacteria in their guts all at once. While this results in a two-day-long bout of itching that is even more excruciating than that in a normal case, sweet relief follows this brief episode. Semanza was fortunate to receive ivermectin in a locally administered program initiated by Katabarwa. Two years after he began treatment the itching was gone, his skin was partially healed, and he was reintegrated with his community, married, and hoping to start a family.[1]

～～～～

Itching can be a brief sensation or it can last for days. In the case of untreated onchocerciasis, it can endure for a lifetime. It can be triggered by mechanical stimuli, like a wool sweater or the subtle movements of an insect's legs over the skin, or by chemical stimuli, like the poison ivy inflammatory agent called urushiol.[2] Itching can also result from damage to sensory nerves or the brain. In some cases it can be triggered by brain tumors, viral infection, or a mental illness like obsessive-compulsive disorder. It's also a well-known side effect of certain therapeutic and recreational drugs.

Itching is highly subject to modulation by cognitive and emotional factors. One night, camping in the Amazon jungle, I was just drifting off to sleep when I felt an itchy sensation on my arm. I got my flashlight and glasses, saw what was causing it, and brushed off a huge millipede. At that point, sleep became impossible. I had become hypervigilant, and every little breeze and twitch evoked a sensation of itch for the rest of the night, and not just on the affected arm but all over my body. I was battling millipedes of the mind until dawn.

The compelling, tormenting nature of itch is well known. In Dante's *Inferno*, falsifiers (including alchemists, impostors, and counterfeiters) were cast into the Eighth Circle of Hell, where they suffered eternal itch (figure 7.1). Only those who committed treachery—fraudulent acts between individuals who shared special bonds of love and trust (like Judas Iscariot, the betrayer of Jesus Christ)—met a presumably worse fate in the Ninth Circle of Hell: being frozen in ice.

Figure 7.1 This 1827 illustration by William Blake shows falsifiers being tormented by eternal itch in the Eighth Circle of Hell in Dante's *Inferno* (Canto 29). The Eighth Circle is actually divided into ten concentric *bolgie*, or ditches, each one presumably worse than the last. The falsifiers occupy the innermost and presumably worst *bolgia*. By comparison, the outermost *bolgia* is for panderers and seducers who are whipped by demons as they march. Used with permission of Harvard Art Museums/Fogg Museum; anonymous loan in honor of Jakob Rosenberg, 63.1979.1.

~~~~~

Here's a question that lies at the intersection of biology and philosophy: Is itch a unique form of touch sensation that is qualitatively different from the other touch modalities, or is it merely a different pattern of stimulation that relies upon one or more of the touch senses we have already encountered in this book? By analogy, is the relationship between itch and other touch sensations like that between a saxophone and a piano? Each produces sound, but those sounds are qualitatively different. Or is it like the relationship between bebop jazz played on the piano and classical music of the Romantic period played on the piano? They, too, are clearly distinguishable because of their musical structure and context, but they come into being on the same sound-producing device. In the past, this type of question would have been left to philosophers. Today, biology can add to the discussion.

Some who believe that itch is a pattern rather than a unique type of touch contend that it is merely a particular type of pain—one of a weak, dilute character. They point out, correctly, that itch and pain have certain similarities. Both can be triggered by a wide variety of stimuli: mechanical, chemical, and sometimes thermal. In particular, both pain and itch can be activated by chemical products of inflammation and can sometimes be relieved by anti-inflammatory drugs. Both are subject to strong modulation by cognitive and emotional factors, including attention, anxiety, and expectation. And both pain and itch signal the intrusion of things in the environment that should be avoided—they are, in other words, motivational sensations that demand action. Pain leads to a reflexive withdrawal response; itch leads to a reflexive scratching response. Scratching in response to itch, like withdrawal from pain to prevent tissue damage, is thought to be protective. It can cause us to dislodge venomous arthropods, like spiders, wasps, or scorpions, or those that transmit disease-causing pathogens, like malarial mosquitoes or plague-bearing fleas.

If itch were merely a weak or intermittent form of pain, then one would imagine that increasing the intensity or frequency of an itchy stimulus could raise it to the threshold of feeling painful, or, conversely, that attenuating a painful stimulus could cause it to evoke an itch sensation. However, when studied in the lab with carefully controlled stimuli, this never happens. Weak pain is just weak pain, and intense itching is just intense itching.[3] Another key distinction between itch and pain involves their location on the body. While pain can be felt widely, in the skin, muscles, joints, and viscera, itching is restricted to the outer layer of the skin (both hairy and glabrous) and the mucous membranes that adjoin the skin, like those that line the mouth, throat, eyes, nose, labia minora, and anus.[4] You can have pain in your guts, but not itchy guts.

If itch is, then, a unique form of touch, then one would expect to find fibers of sensory neurons in the skin that are uniquely activated by itch stimuli and that, when electrically stimulated in the lab, give rise to an itch but not a pain sensation. This is called the labeled-line theory, to distinguish it from the pattern-decoding theory, which holds that the same sensory neurons in the skin can signal either itch or pain, depending upon their electrical firing pattern.[5]

In 1997 Martin Schmelz and his colleagues found the first indications of itch-specific sensory nerve fibers in humans using microneurography, the technique in which a fine electrode is passed through the skin into a sensory nerve to record the electrical activity of single fibers. They found a population of slow, unmyelinated C-fibers that responded electrically when histamine (an itch-inducing chemical that's normally produced in the body) was applied to tiny patches of skin on the legs of volunteers. The electrical response began just as the subjects reported feeling an itch sensation at that same location. Interestingly, these fibers did not target just a small patch of skin but spread to innervate a region about three inches in diameter. Because these fibers did not respond to mechanical stimulation, they were thought to be itch-specific, supporting the labeled-line theory. However, some years later this same group of investigators found that at least some of these itch-responsive C-fibers could also be electrically activated by a pain stimulus, arguing against the labeled-line theory.[6]

Part of the difficulty in interpreting these findings is that the itch stimulus used was histamine, and we know that histamine is only one of a number of different itch triggers that act through different chemical pathways. Indeed, most of us have had the experience of treating an itch with antihistamine cream and finding that it works only in some cases. We don't know from these experiments if the nerve fibers that convey histamine-independent forms of itch are also responsive to pain. And so proof of the existence of labeled-line neurons for itch in humans remains unestablished. An important limitation of these human experiments is that one must hunt blindly with the single-fiber recording electrode: There's no way to see inside a nerve and target a specific fiber. In mice, it's been possible to make a great deal more progress in this area, using genetic, anatomical, and electrical recording techniques.

~~~~~

Many different types of stimulation of the skin can induce itch. In many cases we don't yet have an understanding of the molecular events underlying itch transduction. For most itchy stimuli the pathway appears to be indirect. For example, when skin is repeatedly chafed or responds locally to an allergen, an inflammatory cascade is set in motion (as we discussed in chapter 6—see figure 6.5). Molecules released by immune cells, like histamine from mast cells, can bind to histamine receptors located on the bare endings of sensory neurons in the epidermis and cause them to fire electrical spikes (figure 7.2). In another example, a natural protein fragment called BAM8-22 will bind to a different receptor on itch-conducting nerve endings in the skin, which is called MrgprC11 in mice and hMrgprX1 in humans. Sometimes there can be direct activation of an itch receptor by a molecule in the environment. For example, the antimalarial drug chloroquine is well-known for producing itchy skin. Chloroquine binds directly to a different receptor in sensory neurons called MrgprA3.[7] The key point is that there are at least three different molecular sensors that can activate itch-detecting neurons, and while some are directly activated by signals in the environment, most are responding to a chemical signal in the body that serves as an intermediary.

~~~~~~

If there are truly dedicated labeled-line neurons for itch, then the following should be true: First, one should be able to destroy or silence those neurons and block itch sensation, leaving other touch senses like pain or temperature unaffected. Second, selectively activating those dedicated itch neurons should give rise to a perception of itch, but not pain or other touch sensations. Third, the anatomical distribution of the nerve endings should mirror the known distribution of itch sensation: They should be present in the epidermis of the skin and in external mucous membranes, but not in the muscles, joints, viscera, etc.

One approach to identifying potential labeled-line itch neurons is to try to determine the neurotransmitter molecule used by such itch-detecting sensory neurons to communicate with their targets in the spinal cord and then use genetic manipulation to delete that molecule in a mouse. Santosh Mishra and Mark Hoon of the National Institutes of Health did just that, making an educated guess that the transmitter of itch neurons might be a molecule called NPPB.[8] When they tested an NPPB-lacking mutant mouse, they found that it had severe deficits in itch sensation evoked by a broad range of stimuli, including both histamine and chloroquine. Most important, NPPB-lacking mice had normal responses to pain, temperature, and light touch.

NPPB is released by sensory neuron terminals onto target neurons in the dorsal horn of the spinal cord. Those neurons have receptors that can bind NPPB and then propagate the electrical signal farther along toward the brain. When NPPB was artificially synthesized and injected into the spinal cord of mice, it caused the mice to scratch in a way that looked just like their response to itchy stimuli applied to the skin. After injecting a special toxin into the spinal cord that selectively destroyed neurons containing the NPPB receptor, the mice failed to respond to either itchy stimuli to the skin or to an NPPB injection in the spinal cord. These results suggest that NPPB-using neurons are a labeled line for itch.[9] If that was true, then selectively activating these neurons should produce itch, but not pain or light touch sensation. At the time of this writing, that type of experiment has yet to be reported, but several laboratories are likely attempting it.

The NPPB-expressing neurons that innervate the skin fall into at least two categories. Most of them bear the receptor MrgprA3 on their surface, but some do not (figure 7.2). When the axons of MrgprA3-containing neurons were traced, it was found that they terminated in the epidermis of the skin, but not in the viscera, muscles, or joints, just as we would expect for dedicated itch sensors. When a complicated set of genetic manipulations was performed in mice to enable the experimenters to selectively activate MrgprA3-expressing neurons in the skin, itch responses but not pain responses were observed. (Mice scratch in response to itch but rub in response to pain.) This result argues that MrgprA3-expressing sensory neurons convey itch but not pain information. When MrgprA3-expressing neurons were selectively destroyed, mice had no deficits in pain, temperature, or light touch sensation. However, they did have severe deficits in all forms of itch sensation tested. Importantly, however, the deficits were not total: In particular, significant responses to histamine remained, presumably carried by a population of MrgprA3-lacking itch neurons.[10]

Taken together, the NPPB and MrgprA3 manipulations in mice that we've discussed indicate strongly that there is at least one set of dedicated neurons for itch: NPPB- and MrgprA3-expressing cells. It is possible that there are additional itch-specific neurons as well. It is also probable that there are at least some neurons that convey both pain and itch information and that they may encode these sensations with different patterns of firing. In sum, experimental

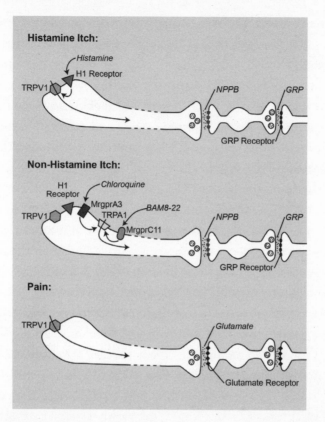

Figure 7.2 Two different C-fiber pathways for itch and their comparison to pain. The neurotransmitter NPPB appears to be specific for itch neurons. In contrast, pain neurons release glutamate to signal to neurons in the dorsal horn of the spinal cord. The spinal neurons that receive NPPB itch signals have NPPB receptors and in turn release a rare neurotransmitter called GRP onto the next neurons in the chain of transmission. Deleting neurons with GRP receptors blocks itch sensation but not pain or light touch, suggesting that this labeled line for itch may be maintained across two synaptic connections.[11] Itch neurons can be divided into at least two categories: those neurons that have the chloroquine receptor MrgprA3 and primarily transduce nonhistamine itch, and those that do not but have histamine receptors and transduce histamine itch. Histamine receptors excite the nerve terminals by opening the ion channel TRPV1, while the receptors for chloroquine and BAM8-22 open the ion channel TRPA1.[12] This diagram is a general scheme only. There are likely to be more populations of itch neurons than those shown here. Also, synaptic interactions between these streams of information in the spinal cord are poorly understood at present.

evidence reveals that there is clearly at least one labeled-line pathway for itch, but we cannot rule out the role of pattern decoding in itch sensation.

What do these results mean for our fundamental neurophilosophical question? The conformation of a labeled line for itch is certainly consistent with itch being unique and qualitatively distinct. That said, we don't yet know what happens to that stream of itch information on the way to the brain. It is almost certainly blended with other touch sensations to some degree, potentially degrading its specificity. In the end, perhaps the clearest guide is our human experience: There is a unique word for itch in almost every language studied to date.

On the practical side, identification of itch-specific receptors and neurotransmitters opens the door to new itch-reducing therapies. In the future it may be that choloroquine treatment for malaria will be given together with a second drug that blocks MrgprA3. Because many types of itch are not effectively relieved by antihistamines or other anti-inflammatory drugs (like steroids), they might be effectively treated by novel drugs that selectively block MrgprC11 or NPPB receptors or GRP receptors. As always in drug development, there will be hurdles. For example, NPPB has an important signaling function in the heart, and so NPPB receptor antagonists might have cardiac side effects that make them unsuitable as itch-controlling therapies.

> *"Happiness is having a scratch for every itch."*
> —Ogden Nash

It feels so good to scratch an itch. Even when we know that the itch will return even stronger after the scratching has ceased, most of us can't help but do it repeatedly. Itch is so compelling, and the relief of itch is so satisfying, that *itching* is now used in everyday speech to mean "strong urge." In the lyrics of the song "Hesitating

Beauty," Woody Guthrie wrote, "Well I know that you are itching to get married, Nora Lee / And I know I am twitching for the same thing, Nora Lee." We understand precisely what he means, as *itch* is an apt metaphor for unconsummated longing.[13]

In one unpleasant experiment, volunteers were treated with spicules of the cowhage plant, which induce an intense itch sensation. Cowhage spicules were applied either to the forearm, the ankle, or the back, and these sites were then scratched by the experimenter using a small brush. Every thirty seconds the subjects had to rate the intensity of the itch and the pleasure produced by scratching it. What the experimenters found was that scratching the back was the most effective in reducing the perception of itch but that scratching the ankle was the most pleasurable sensation.[14]

Why does scratching temporarily relieve itch sensation? We don't entirely know. One theory holds that our perception of itch is dependent upon a balance of itch and pain signals that converge at some location in the spinal cord; when we scratch, we produce mild pain that competes with the itch sensation and thereby reduces it. Pain from a pinprick, electrical stimulation, or noxious heat or cold can also serve to relieve itch. However, it seems that in some cases even light scratching that is below the threshold for pain can relieve itch.

A related theory holds that when a very localized stimulus on the skin occurs, like that produced by the legs of a small insect, it can activate itch sensory neurons, and that this local sensation passes through the spinal cord to the brain unmolested to evoke itch sensation. However, when that region is subsequently scratched, activating touch sensors over a wide area of skin, it engages inhibitory circuitry in the spinal cord that attenuates the passage of itch sensation to the brain.[15] It may be that we have been wired by evolution to have small, localized skin sensations feel itchy in order to evoke the scratching reflex and reduce our exposure to insect-borne toxins and infection.[16]

It's well-known that opiates like heroin and oxycodone can trigger intense bouts of itching. Heroin users will often praise a particularly itchy batch of the drug, based upon the correct idea that itchiness and psychoactive potency are related. Drug counselors and probation officers have learned to be alert for scratching as a sign of chronic opiate use. Opiate-triggered itch also occurs in a clinical setting: About 80 percent of patients treated with opiates for pain relief experience itching, which occurs even if the opiate is injected directly into the fluid surrounding the spinal cord, minimizing its direct effects on the brain and the sensory nerves.

For many years it was thought that opiate-triggered itching was a secondary effect of pain relief. The underlying concept was that if pain and itch signals converge and compete in the spinal cord, then blocking pain signals will tilt the balance of power toward itch. This completely reasonable hypothesis turned out to be dead wrong. Opiate-induced analgesia and opiate-induced itching are completely separate phenomena. One set of neurons located in layer II of the spinal cord dorsal horn receives pain signals and expresses the mu-opioid receptor. When opiates like heroin bind the mu-opioid receptor, the propagation of electrical signals to the brain is blocked, thereby producing pain relief. (Opiates mimic the natural action of endorphins released by the descending pain-modulatory system.) A separate set of neurons in the spinal cord resides in spinal dorsal horn layer I and receives itch signals relayed by the neurotransmitter GRP. These itch-signaling neurons express a hybrid receptor in which one part is a receptor for GRP and the other part is a special type of the mu-opioid receptor called MOR1D. When you take an opiate, the analgesia and itching are produced simultaneously but are mediated by separate receptor molecules and separate neural circuits in the spinal cord. The good news is that it will likely be possible to produce a combination of therapeutic drugs or a new derivative of morphine that would relieve pain without producing itch.[17]

Just as there is no single location in the brain for pain, no one brain region is responsible for the itch sensation. Indeed, in a general, low-resolution sense, the brain regions activated by itch and pain are similar. Itch activates both sensory-discriminative regions, like the thalamus and the primary and secondary somatosensory cortices, and more affective-emotional-cognitive regions, like the amygdala, insula, anterior cingulate, and prefrontal cortex. (For the pain circuit, see figure 6.4.) Both pain and itch also ultimately activate regions involved in motor planning and coordination to produce and modulate withdrawal and scratching responses, respectively.[18]

There are diverse forms of brain and spinal cord damage that can result in chronic itch. These include traumatic injury, certain types of stroke, brain tumors, infections, multiple sclerosis, and other autoimmune disorders. Chronic itch can also result from damage to the sensory nerves (from compression, trauma, tumors, diabetes, or infection).[19] As in the case of pain, the interactions between the sensory nerves and the brain are complex, and when the sensory nerves conveying itch are damaged, aberrant itch signals are propagated to the brain. The continued bombardment of the brain by these signals (or the absence of normal itch signals) can ultimately rewire the brain's itch circuits. People with amputated limbs, for example, can have phantom itch as well as phantom pain.

One of the most common forms of itch that results from damage to neurons follows the disease called shingles. Shingles, a common infection of sensory nerves by the herpes zoster virus, is characterized by a painful rash, most often affecting the elderly or those with compromised immune systems. Shingles destroys sensory neurons and, particularly when it occurs on the head, can ultimately result in a chronic maddening itch that cannot be effectively treated with antihistamines or steroids.

The most dramatic case of itch following shingles in the medical literature is that of M., a thirty-nine-year-old woman who developed shingles and was successfully treated with an antiviral drug.[20] Shortly thereafter she felt itching and numbness on her right forehead, which she began to scratch with her fingernails. When M. sought medical attention for the problem, her internist found no infection or allergic reaction in her skin, and the itch was not relieved by anti-inflammatory creams. She was then told that her itch was psychiatric in origin—the result of depression and an obsessive-compulsive disorder. However, treatment with drugs for those conditions also failed to alleviate her itch. During the day M. could sometimes resist the urge to scratch, but at night she would scratch furiously during sleep. She tried wearing caps to bed but in the morning would always find the cap removed and blood on her pillow. With time she scratched away a patch of her hair at the itchy spot, and a scab formed. One morning ten months after her shingles had resolved, she was horrified to find a foul greenish liquid running down her face.

M. arrived at the emergency department at Massachusetts General Hospital, where it was soon determined that she had scratched through skin and skull to reach her now-exposed brain (figure 7.3); the greenish liquid was her own cerebrospinal fluid. The surgeons formed a skin graft to cover the wound, but M. soon scratched it away in her sleep. She was then outfitted with a foam helmet and mitts secured to her wrists with strong tape, to wear at bedtime. Psychiatric testing conducted at this time revealed that she did not in fact suffer from obsessive-compulsive disorder or hallucinate, yet her uncontrollable scratching made her a danger to herself. She spent two years confined to a locked medical ward, and today, years later, lives independently. She has been able to control her nocturnal scratching and has adopted coping strategies: If she must scratch, she'll do it gently with a soft, rolled-up cloth. The brain damage from her scratching has left her partially paralyzed on the left side and has changed her personality, as often occurs with frontal-lobe damage.

The neurological origin of M.'s terrible postshingles itch is not

Figure 7.3 Following an episode of shingles, this thirty-nine-year-old woman developed a terrible, unrelenting itch on her right forehead. The shingles had left her skin numb in that location, so she felt no pain as she scratched her way through skin and skull, ultimately contacting her brain. The bottom image shows a CT scan revealing the hole in the skull with the right frontal lobe protruding and damage to the underlying brain tissue. From A. L. Oaklander, S. P. Cohen, and S. V. Y. Raju, "Intractable postherpetic itch and cutaneous deafferentation after facial shingles," *Pain* 96 (2002): 9–12, with permission of Elsevier.

entirely clear. Her light-touch, temperature, and pain and itch perception are normal on all parts of her skin other than the chroni-
cally itchy spot on her right forehead. That particular region has
severe tactile deficits, with little perception of light touch, temperature, or pain—which explains how she could scratch through her
skin without being overwhelmed by pain. A tissue biopsy revealed
that about 96 percent of the sensory nerves in that patch of skin
had died, yet application of an anesthetic gel that blocks neural firing was able to provide temporary relief of the itch, suggesting that
the few surviving nerve fibers were sufficient to trigger the unrelenting itch sensation. (Conversely, those same nerve fibers could
also convey signals in response to scratching that would relieve the
itch.)

In a sense, the portion of M.'s right forehead that has lost most
of its nerve fibers is a lot like an amputated limb: The brain is receiving sparse and atypical information from a bodily region and is
trying to interpret those signals. It's also likely that the brain has
rewired itself in response to the nerve damage and that the rewiring is involved in the chronic itch. The itch cannot result solely
from activity in the itch centers of the brain, since blocking signals
in the remaining forehead sensory nerves can provide temporary
relief. It's also unlikely that the brain is functioning entirely normally, and that the chronic itch is purely a result of aberrant signals
in the sensory nerves. The most likely explanation is that the signals from the remaining sensory nerves trigger aberrant activity in
the brain's itch centers to create the hellish, unrelenting itch.

When people showed up for a free public lecture in the German
university town of Giessen, they didn't realize that they were to be
the subjects of an unusual experiment. The title of the lecture, presented in cooperation with a public television station, was "Itching—
What's Behind It?" Video cameras in the hall were trained on the
audience as well as on the speaker. The aim of the experiment was

to determine if the itch sensation could be induced in the audience by showing pictures of fleas, mites, scratch marks on skin, and skin rashes. As a control, images of bathers and mothers with infants were also presented (soft, hydrated skin suggesting absence of itch). It's not surprising that a significant increase in audience-member scratching frequency was induced by itch-related images.[21] Subsequent experiments in a laboratory setting using itch-themed videos have confirmed this basic finding and have shown that the subjects need not be suffering from a preexisting skin condition in order to experience this socially contagious itch.[22] One interesting proposal to explain this phenomenon was that people who were more empathetic were more likely to feel itchy themselves when they observed another person scratching. However, when personality questionnaires were given to subjects in these experiments, no correlation between empathy and social itch contagion was found. Instead, people with the greatest tendency to experience negative emotions (high neuroticism) were most likely to be subject to social itch contagion.[23]

Why is it that watching someone else's finger being hit with a hammer will not usually make us withdraw our own fingers, but watching someone scratch will make us feel itchy and cause us to scratch as well? The best guess is as follows: Through most of our human history we have been routinely exposed to disease- and toxin-bearing parasites. In situations where these occur, if you notice that the person next to you is scratching, there is a good reason to believe that you are also being exposed to the same dangerous insect, worm, etc., and it's therefore adaptive for you to feel itchy and scratch in order to reduce your own chance of harm. Pain, in contrast, is weakly socially contagious, because the cause of most pain is not generally spread from person to person.

Imagine that you are in a subway car and the person sitting next to you begins to scratch uncontrollably. That stranger is clearly tormented but—be honest now—is your first reaction compassion or revulsion? André Gide examines this question:

The itch from which I have suffered for months . . . has recently become unbearable and, for the last few nights, has almost completely kept me from sleeping. I think of Job looking for a piece of glass with which to scratch himself and of Flaubert, whose correspondence in the last part of his life speaks of similar itchings. I tell myself that each of us has his sufferings, and that it would be most unwise to long to change them; but I believe that a real pain would take less of my attention and would after all be more bearable. And, in the scale of sufferings, a real pain is something nobler, more august; the itch is a mean, unconfessable, ridiculous malady; one can pity someone who is suffering; someone who wants to scratch himself makes one laugh.[24]

Unrelenting itch may indeed be the worst form of sensory torment. Perhaps Dante should have reserved it for the worst sinners in the innermost Circle of Hell. The compulsion to scratch is overwhelming, and yet when we do so, those around us recoil and regard us as doubly cursed: both infested and weak willed.

CHAPTER EIGHT

ILLUSION AND TRANSCENDENCE

The real tactile world is wonderfully messy and complicated. It's not composed of easily separable touch stimuli—a little pain on the abdomen here, followed by a gentle caress on the arm there. In the early part of the twentieth century, psychologists studying perception began to recognize that many of the most important and motivating tactile sensations, such as wetness, greasiness, or stickiness, might not be basic touch senses with dedicated detectors in the skin, but rather what they came to call touch blends. These old scientific papers make for delightful reading. In experiments designed to explore the factors that result in a sensation of "clamminess," M. J. Zigler of Princeton University used a series of increasingly creepy stimuli. Writing in 1923, he quoted an earlier researcher who claimed that a "horribly clammy feel is touched off when the finger is set upon by a bit of cold boiled potato in the dark." In his own work he found that a cold boiled potato was in-sufficiently clammy, so he searched for a more effective touch stim-ulus, eventually settling on a damp kid glove filled with moistened oatmeal. Zigler concluded that clamminess is a blend of touch sen-sations, most notably coldness and yielding softness, and went on to state that "a genuine clammy experience" will always be per-ceived as unpleasant.[1]

To study the perception of wetness through touch, I. M. Bentley of Cornell University first sought to eliminate the other senses. His subjects were blindfolded, and their ears and noses were stuffed with cotton. Their right hands were secured, palm down, with their extended middle fingers hanging over the edge of the table. Test liquids, consisting of mercury, petroleum, water, Eldorado oil[2], molasses, benzene, and ether were poured into a beaker, which was carefully raised, using a system of pulleys, to submerge the middle finger. Bentley argued, "The apprehension of wetness is commonly considered something unique: The finger touches a wet surface, or the hand is plunged into a liquid, or the body is immersed, and one is said just 'to feel wet.' This is a striking instance of the confusion of mental processes and their significance; in this case, a confusion of sensation with perception." He concluded, after many pages, that the perception of wetness was likewise a touch blend, most strongly influenced by temperature and pressure.[3]

Were Bentley and Zigler correct about the blended nature of the respective complex touch perceptions they studied, or should we expect that unique wetness- or clamminess- (or stickiness- or greasiness-) detecting molecules in dedicated sensory neurons will be found someday? The answer is almost certainly the former. Neither experiments on human perception, recordings of single sensory nerve fibers, nor molecular genetic studies are pointing to dedicated sensory neurons for these sensations.

You may be thinking that this consideration of the validity of the existence of touch blends is a straw man argument—after all, could we really have expected that there would be dedicated sensory neurons for sensations like wetness or clamminess? But we should be humbled by the lessons of our past experience. It is only very recently that we have obtained strong evidence for unique itch receptors, and it would have been reasonable before then to maintain that itch, too, is always experienced as a touch blend. That's the fundamental power of biology: No amount of philosophical

reasoning, linguistic analysis, or introspection could ever have resolved such matters.

~~~~~~

So what have we learned about touch? Starting at the skin, we have
an assortment of nerve endings ranging from bare nerves to specialized structures with wild and elaborate shapes. Each of these is
a molecular machine fine-tuned by evolution to extract different
aspects of information about our tactile world. The nerve fibers
that relay the information from touch sensors in the skin to the
spinal cord are mostly (but not entirely) dedicated to a single class
of sensor for mechanical sensation: one tuned for rough texture,
another for vibration, another for stretching. Surprisingly, we even
have specialized sensors for caress, itch, and, quite likely, sexual
touch. Some of these streams of information are conveyed by fast
fibers, and some by slow ones, so they arrive at the brain with different delays. Most often, slowly transmitted information, like that
from the caress sensors or the second wave of pain, activates the
emotional-affective-cognitive portions of the brain's touch circuits, while fast information, like that from traditional mechanosensors, activates sensory-discriminative centers. The streams of
touch information from these various dedicated detectors are combined with signals about attention, emotional state, and past history so that by the time we have conscious access to touch sensation,
it is in the form of a unified and useful percept with both discriminative and emotional properties.

Crucially, our brains are not passive recipients of touch information; rather, they can send descending signals to the spinal cord
to turn the gain of touch signals up or down before they reach the
brain. This has been most easily studied in the descending pain
system, but is likely to be relevant for other aspects of touch as well.
The touch circuits of the brain and spinal cord have been sculpted
to solve particular evolutionary problems: how to find food, avoid

danger, mate, protect offspring, etc. Indeed, all touch sensation (or sensation of any kind) is ultimately in the service of action. Our touch circuits are not built to be faithful reporters of the outside world but are constructed to make inferences about the tactile world based upon expectations—expectations derived from both the historical experience of our human ancestors and from our own individual experiences. And finally, we've learned that interpersonal touch not only has a special role in early human development but continues to be crucial across the span of human social life, promoting trust and cooperation and thereby deeply influencing our perceptions of others.

All these discoveries about touch are significant, for they help us understand a central aspect of our human experience. But at this point in time our overall examination of touch has been rather limited. We need to get even messier than a kid glove full of wet oatmeal and expand our study of the tactile world. We need to explore not only blended touch sensations but also the effects of touch and nontouch senses combined. And finally, we ought to move beyond simple touch sensations to investigating phenomena like illusions, everyday hallucinations, and those transcendent touch experiences that, at first glance, seem to require a supernatural explanation.

~~~~~~

At about the time they were three years old, my twins, Natalie and Jacob, began to play a game that would inevitably end with tears and injury. Each would stand on one side of the bathroom door, and they would take turns pushing it toward each other and laughing. The kids got along well and never engaged in conventional shoving matches, but they loved the bathroom door game. This activity was fun because each child was blocked from the other's view, giving the door the illusion of agency, as if it had become animated with its own life force, like the objects in their favorite cartoons.

Although the game began with gentle pushes, it would inevitably escalate, each push becoming more forceful, until one child would get smacked in the head with the door and a parent would have to calm everyone down.

"Natalie, why did you push the door into Jacob's face?"

"It's not my fault. We were taking turns, but each time he pushed harder, so then I would push back the way he did."

"No, Nallie!" Jacob would interject (at age three, he couldn't quite pronounce her name). "*You're* the one who pushed harder, not me."

"No way! It was *you!*"

"Kids," I told them, "I don't want you playing the bathroom door game. Someone always gets hurt."

"Okay, Daddy," they would say in unison, their promise almost instantly beginning to fade from their minds.

While there are social factors involved in this escalating-force situation, the central explanation for the phenomenon actually comes from the neurobiology of touch processing. We are wired to pay less attention to touch signals that result from our own movements as compared to those that originate in the outside world. For example, when we walk down the street, we barely notice the sensations of our clothing moving against our skin. However, if we experienced these identical sensations while we were standing still, they would be very conspicuous and would demand our immediate attention: Who or what is rubbing up against us? This makes sense: Externally generated sensations are the ones that are most likely to demand our attention because they are potentially threatening or otherwise salient (flirtatious, delicious, puzzling, etc.).

In the bathroom door game Natalie and Jacob were each trying to match the other's pushing force in alternate turns. However, this is an almost impossible task. When Natalie pushed with two units of force, Jacob felt two units of force on his extended palms. Yet when he tried to match her with an equal amount, he exerted three units of force. Why? Because three units of self-generated force gave the same feeling of pressure on the skin of his palms as two units of force produced by his sister. Then Natalie felt three units of force from Jacob, and in her attempt to match that, she applied four units, and off they went to Armageddon.[4]

The perceived attenuation of self-generated touch is seen in many different tactile situations. As discussed in chapter 6, it's been shown that self-inflicted pain is rated as less intense and less unpleasant than the same pain administered by another person or by stimuli that are randomly triggered by a computer. And, of course, partnered sex, in which the other person is moving independently, feels fundamentally different from self-pleasuring, where your brain can predict your own stimulating movements (even if you're using a mechanical device like a vibrator).

What's the neural basis of attenuated sensation during self-touching? One of the clearest answers comes from the study of tickling. Most people can't tickle themselves effectively; the tactile sensation from self-tickling is much weaker than that which results from being tickled by another person. Sarah-Jayne Blakemore and her colleagues at the Institute of Neurology in London have performed experiments in which subjects were either tickled or instructed to self-tickle while they were in a brain-scanning machine. The experiment was carefully controlled so that the location, force, and pattern of the tickling were the same in the tickling and self-tickling cases. Tickling resulted in activation of both the sensory-discriminative portions of the touch circuit, such as the primary and secondary somatosensory cortices, as well as certain regions in the emotional-affective-cognitive touch circuit, such as the anterior cingulate. When the experiment was repeated with self-tickling, activation of the brain's touch centers was reduced when compared with conventional tickling. At the same time, self-tickling produced a strong stimulation of the cerebellum, a brain structure that receives both touch signals and instructions from other brain regions that initiate motion, like the electrical signals that flow through neurons to control muscles in the hand and arm during self-tickling. The cerebellum is activated when instructions for motion are specifically correlated with sensory feedback from the skin's touch sensors. It then sends signals to the brain's touch centers (and other regions) that dampen that activation and thereby attenuate the ticklish sensation during self-tickling.[5]

The failure of self-tickling is dependent upon close correlation between the tickling movement and the touch sensation. When a computerized mechanical tickler[6] was interposed between the self-tickling hand and the skin such that the movements of the hand were translated into movements on the skin, not immediately, as would occur with natural self-tickling, but rather with a delay of 200 milliseconds, then the self-tickling sensation became much stronger. When the direction of the tickling was altered by the computer (changing up and down to side to side, for example), it produced further increases in the efficacy of self-tickling. Brain imaging in those conditions revealed that as the commands for movement were uncoupled from the skin sensations on the tickled body part, cerebellar activity was reduced, resulting in increased activation of the brain's touch circuit and increased ticklishness.[7] Interestingly, some people with cerebellar damage can tickle themselves effectively, as can some schizophrenics (who may also have cerebellar dysfunction). It may be that in both of these conditions the cerebellar circuits, which function to compare commands for motion with sensory feedback, are malfunctioning. The result of this malfunction is that self-generated touch sensations feel as if they were coming from the outside world.[8]

The tactile confusion that occurs in the cases of schizophrenia or cerebellar damage is reminiscent of another state that all of us have experienced: During dreaming sleep, we often hallucinate and are confused about the origins of sensation. In one report women who had just been awakened from the REM stage of sleep (the stage when storylike dreaming is most likely to occur) reported a greater response to self-tickling than to conventional tickling.[9] Perhaps the relevant circuits for dialogue between the cortex and the cerebellum are suppressed during REM sleep and take time to come back online after waking. During REM sleep, commands flow from the brain's motor-control centers, but they are actively blocked so that they cannot reach the spinal cord and ultimately control the muscles. Because of this blockade the body goes almost totally limp, and even the muscles that are normally contracted during other stages of sleep become relaxed. For that reason you can't maintain REM sleep while sitting in a chair—you'll eventually slump over and awaken.[10]

The flaccid paralysis that occurs during REM sleep gives rise to another phenomenon. While most people recover voluntary control of their muscles quickly upon waking from REM sleep, some

Figure 8.1 A male sleep demon, called an incubus, applies his weight to the abdomen of a sleeper. In many traditions it is believed that sexual intercourse with an incubus can result in ill health or even death. The folk tale of the incubus (and its female counterpart, the succubus) may derive in part from the phenomenon of sleep paralysis, in which waking from a narrative dream is accompanied by the sensation of a crushing weight. This etching, called *The Covent Garden Night Mare*, was made by Thomas Rowlandson in 1784. It is a satiric adaptation of a well-known painting, *The Nightmare* by Henry Fuseli (1781). The original version features an attractive young woman, replaced here by the British politician Charles James Fox. Used with permission of the City of Westminster Archives Center.

experience a period of tens of seconds or even minutes between the moment they awaken from REM and the time when the blockade of motor signals to the spinal cord is finally relieved. In this situation, people are awake but temporarily paralyzed. When their commands to move are generated, there are no corresponding touch or proprioceptive feedback signals, so the brain makes the inference that there must be a crushing weight on the body that prevents movement. This terrifying tactile hallucination is fairly common[11] and has likely influenced folk tales of chest- and abdomen-compressing malevolent sleep demons that are found in many cultures (figure 8.1).

~~~~~~

Sensory illusions can also be fun and can provide a glimpse behind the curtain to reveal the brain's subconscious perceptual strategies. There are many illusions in the domain of touch, and my favorite is called the cutaneous rabbit.[12] If you were to close your eyes and then I were to give six quick taps to your inner forearm with equal time intervals between the taps, delivering the first three to your wrist and the second three to your inner elbow, you would feel the first tap at your wrist, but the subsequent five would appear to hop along the intervening unstimulated skin in the direction of your elbow (figure 8.2). We know that this effect is produced in the brain, not by some mechanical effect on the skin, because the rabbit hops are felt even when the intervening area of skin is anesthetized. Furthermore, when the cutaneous rabbit illusion is performed on a subject in a brain scanner, the pattern of activation matches the illusory experience: Hopping activation in the intervening sites along the map of the arm is observed in the somatosensory cortex.[13]

While we don't fully understand the neural basis of the cutaneous rabbit illusion, the best overall hypothesis involves expectation. Either through our individual experience in the world or through the genetically encoded experience of the human species, we have come to expect that tapping stimuli typically progress slowly along the skin. As a consequence, when there is a large spatial gap between

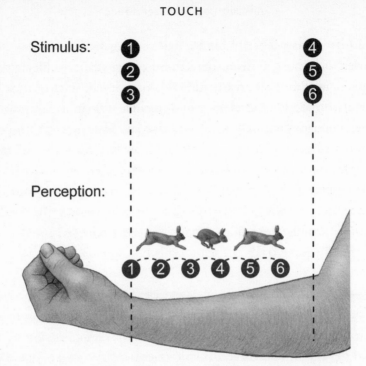

Figure 8.2 The cutaneous rabbit illusion. Three taps at the wrist followed quickly by three taps at the elbow are perceived as a tap at the wrist followed by progressive hops of the tap stimulus along the forearm.

taps three and four, the sensation of taps four through six is blended with the expectation that these types of signals move slowly. The resultant perception of a tap at an intervening location between elbow and wrist is the brain's inference: a sort of compromise between the touch signals it receives from skin at the elbow and a prior expectation of slow movement. What's even odder is that the perception of taps two and three are also altered. This effect is called postdiction, and it relies upon the short delay (about 0.2 second) between the time the tap is received on the skin and when it is perceived. During that perceptual delay, the brain can modulate touch processing based upon the continuing inflow of information blended with expectation to reach back in time and change the perception of events that just occurred.[14]

Our brains naturally bind together information from multiple senses to create a holistic perception of events and objects. Prior expectation can also create illusions when touch sensation is coincident with another sensory stimulus, such as sound. For example, when you rub your palms together like an old-fashioned movie villain plotting a caper, you are simultaneously feeling the sensations on your palmar skin and hearing the whispery noise that results. In one clever experiment subjects were instructed to rub their palms together while a nearby microphone picked up the resultant sounds, which were played back to the subject through headphones. In some trials the sound through the headphones was unaltered, but in others it was modified to boost the high-frequency audio components, which made the subjects perceive their skin as smoother and drier—more like paper. Hence, the name of this effect: the parchment-skin illusion. When an electronic circuit was used to delay the audio feedback by one-tenth of a second, the illusion failed, and the subjects' skin felt normal. In order for the brain to translate high-frequency sound into the perception of smoother, drier skin, the sound must appear to result from the action of the palms rubbing together.[15]

In 1846 Ernst Weber, one of the founders of modern psychology, reported that a large cold coin (a Prussian silver thaler) placed on the forehead feels much heavier than a warm one. It's a significant effect: Most subjects in the experiment perceived the cold coin as being fourfold heavier than the warm one. A similar result occurred when the coin was placed on the forearm.[16] Generally speaking, we don't automatically expect cold objects to be heavier, so the most likely explanation for the effect lies in the sensory nerves, not the brain. Many of the Merkel disk sensory neurons (which respond persistently to pressure) are also activated by sudden cooling of the skin,[17] and it is this activation that is likely to underlie the temperature-weight illusion. The take-home message: Not all tactile illusions result from prior expectations (or any process in the brain at all). They can also arise simply from the tuning of sensory neurons in the skin.

~~~~~

Now let's move from the illusory to the transcendent. A few months ago I was holding Z. and her skin felt wonderful—not just soft and warm, but electric and glowing. At one point we were touching each other on the arms, neck, and back, and the sensation was truly amazing: We could both feel the electrical hum of loving attraction. A few minutes later, while continuing to touch and talk, we had a minor disagreement, a mere ripple in an otherwise tranquil and sweet interlude. The slight rift ended after only a few minutes, and when we were back to feeling close, she asked, "Did you notice how different our touching felt as soon as the mood changed? It was something subtle about skin texture that changed the touch from electric to ordinary. And then later, when it was resolved, did you feel that electric vibe come back?" I had indeed.

Nearly everyone has experienced that wonderful, connected sensation of loving touch. How can we understand it in terms of neurobiology? It's not just a matter of such sensory properties as soft, warm, and yielding. After all, cuddling with your cat is delightful, and it can sometimes be even softer and warmer than your human partner, but that romantic glow just isn't there.[18] Certainly a large part of the loving-touch sensation can be explained by emotional and cognitive modulation of touch perception within the brain. We've discussed how these states can affect perception of pain, so it's not too surprising that they can also affect other touch sensations. But the loving-touch vibe cannot be explained solely by events in the brain.

In addition to the sensory nerves carrying touch information to the brain, there are autonomic nerve fibers that enable the brain to actually alter the properties of the skin. One's emotional state can produce subconscious activation of the autonomic nervous system to affect sweating and local blood flow and erect the hairs on our skin, particularly on the arms. (Of course these emotionally driven skin changes are accompanied by other bodily changes: respiration, heart rate, core muscle tone, and so on.) When our emotional state changes our skin—say, by making us sweat or erecting hairs—it has reverberating effects on interpersonal touch. The caress and other sensors that are activated by erect hairs encode movements differently from ones that lie flat. Likewise, a layer of sweat will affect the way your texture and pressure sensors will be activated as you explore your partner's skin. These subconscious changes in skin properties will cause your skin to feel different to your partner as well, and your own emotional state will be modulated by your perception of your partner's reaction to touching you. It's not just a meeting of minds or a meeting of skins, but a dual dialogue between mind and skin that, in the best of cases, reverberates from one body to the other in a positive way.[19]

~~~~~

Our body schema—the brain's internal map of our body in space—can expand and morph to encompass inanimate objects that we touch and control. This explains why we instinctively duck our heads when driving a vehicle that barely clears an overpass, and why Texas politicians sporting cowboy hats duck their heads when going through Capitol doorways—their hats have become automatic extensions of their bodies. Similarly, the body schema of a ditch digger comes to include her shovel, and a violinist, her bow, each of which can come to function as a tactile sensory appendage.[20] But these useful yet odd effects on body schema are not limited to things that we actually touch. We can respond to sensory stimuli that don't impinge on our bodies at all.

When my children were small, they would delight in being tickled again and again. Soon I came to learn what all parents know: Once your kids have been tickled a few times and are all wound up, you don't even have to touch them to make them convulse with ticklish laughter. Wiggling your fingers a few inches above the ribs will usually do the trick. The effect works even better if you make a noise that becomes associated with the tickling (I would make a high-pitched whine like the sound of a hummingbird) and then repeat that noise while fake tickling. While most adults lose this sensitivity to fake tickling, some retain it into adulthood.

The modern adult equivalent of fake tickling is the phantom cell phone vibration. In a recent survey of medical staff at an academic medical center in Massachusetts, 68 percent of cell phone users reported sometimes feeling the sensation of a vibrating cell phone when in fact the phone was not actually vibrating—or, in some cases, not even being carried. Thirteen percent of the survey's respondents reported feeling phantom cell phone vibrations at least once per day. Those who carried the phone in a breast pocket were more likely to experience phantom vibrations than those who used a belt clip.[21] While fake tickling is a nontouch stimulus (involving sight and sound) that is perceived as touch, phantom cell phone vibration results from no stimulus at all and hence qualifies as a full-blown hallucination. Both fake tickling and phantom cell phone vibrations are the products of expectation based on prior individual experience. If these phenomena were examined in the brain scanner, it's likely that both would show activation of the appropriate site in the body map of the primary somatosensory cortex.

> *"By the pricking of my thumbs,*
> *Something wicked this way comes."*
> —William Shakespeare, *Macbeth*, act 4, scene 1,
> spoken by the Second Witch[22]

People love to talk about mysterious tactile sensations, much like the foreboding of Shakespeare's witch.

"My grandpa can feel changes in the weather coming with his arthritic knee."

"I always have an itchy feeling on my neck before bad news arrives."

Such claims are typically delivered in an awed (or sometimes emphatic) tone of voice. The implication is that these are phenomena that cannot be explained solely by understanding the natural world but require supernatural explanations of one form or another. In the case of the arthritic knee, one could imagine that changes in barometric pressure that precede weather events might subtly alter the conformation of tissues in the knee, thereby providing a naturalistic explanation. However, the bulk of evidence is weighed against this idea: Despite widely held beliefs[23] that have persisted since the time of Hippocrates (approximately 400 BC), research has not established a clear association between arthritis pain and the weather.[24] In the case of the prescient itchy neck, the most likely explanation is recollection bias: The times when an itchy neck was not followed by bad news are forgotten, and only the confirming cases are recalled, leading to a spurious correlation.

As you can tell, like most scientists, I'm skeptical about such claims. But only a week ago I was in a restaurant when I felt a weird prickly sensation on the back of my head, which gave me the strong sensation of being watched. I turned around and, sure enough, two booths away, an elderly couple was looking directly at me. I waved, their eyes darted down to their plates, and life went on. We've all experienced the tactile feeling of being watched. Certainly recollection bias plays a role here. I tend to remember the times when I turned around to find myself being watched but fail to recall the times when I turned around to find no one looking at me. That said, I don't believe that recollection bias is a complete explanation of the phenomenon.

It's been well established that when deprived of all relevant sensory information in the laboratory, we cannot detect someone's presence behind us, much less their gaze. But in the real world, we're not subject to such extreme deprivation. We can detect objects and motion at the edges of our visual field that do not enter our conscious perception. When an event is completely outside our visual field, we often sense other cues, like the sound of halted or hushed conversation or changes in air pressure when the door to a room is opened. Crucially, we do not need to be consciously attending to these sensory cues for them to affect our perception. Just as with the phantom cell phone vibration, when we hear an ongoing conversation stop or change rhythm or volume behind us, or when we feel the subtle breeze that accompanies a door opening, our brain makes an inference based upon past experience and creates a tactile sensation where none exists. In my case, such a signal accounted for the tingling I felt at the back of my head.

When confronted with experiences that are deeply emotional, transporting, or counterintuitive, it is a fundamentally human response to seek explanations beyond the natural world. Touch is intrinsically emotional, and so the experience of touch is often subject to such inclinations. Yet the supernatural is not required to explain mysterious or transcendent touch sensations.

Whether it involves the electric touch of romantic love, the unsettling feeling of being watched, the relief of pain from mindful practice, or the essential touch that newborns need to thrive and communities need to cohere, the transcendent aspect of touch prevails when we understand that these feelings flow from the evolved nature of our skin, nerves, and brain. Ultimately, the biology of touch shows us that the natural is as deeply human and humane as the supernatural.

ACKNOWLEDGMENTS

A reader will ask, "David, in your work on the sense of touch, do you find that—?"

And I sheepishly interject, "Um, well, actually . . ."

It's a reasonable inference. I'm a brain researcher and I wrote a book about touch, so it's natural to assume that I study the biology of touch in my lab. Actually, that's not the case. My lab works on lots of cool things ranging from memory to exercise to recovery of function after brain injury. But I'm not a touch researcher. Rather, I'm an ambassador from the nation of neuroscience, here to pass along the good word from the deep interior of the continent. In fact, ambassador is too lofty a title. Really, I wrote this book because over the years I have become an unabashed *fanboy* of touch research, starting at my academic home, the Johns Hopkins University School of Medicine. I first became intrigued by the pioneering work of Steven Hsiao and Kenneth Johnson (both disciples of the legendary touch researcher Vernon Mountcastle) on the neural basis of texture and form discrimination. Then the beautiful and rigorous experiments of David Ginty and his colleagues on the molecular identification of fine touch receptors opened a whole new level of analysis of tactile sensation, as did the molecular and genetic insights of Michael Caterina and Xinzhong Dong on the topics of pain, temperature, and itch. And recently, work of the young scientist Daniel O'Connor seeks to bridge the gap between touch sense and decisions. I am indebted to the home team, these brilliant, creative, and kind researchers, for firing me up about this fascinating topic. I've also received lots of encouragement and advice from the broader community of touch researchers around the world. Thanks to all who took the time to chat in the hallways of meetings or answer late-night e-mails.

It's always fun to have some experts in your corner. Steven Hsiao, Sascha du Lac, Daniel O'Connor, Michael Caterina, David Ginty, and Xinzhong Dong went above and beyond the call of duty to engage carefully and constructively with my early manuscript drafts. I am also indebted to my lay readers who pushed for greater clarity, let me know when things got too nerdy, and rightfully insisted that I cut some of the lamest and most offensive anecdotes (they can't be blamed for the ones that remain). A huge shout-out is due Marion Winik, Kate Sanford, John Lane, and Laura Coleson-Schreur.

I give a deep bow to the publishing pros. Once again, Joan Tycko has contributed lovely and clear illustrations. Rick Kot, a gentleman among editors, has wielded his scalpel, retractor, and probe with grace and sensitivity. Rimjhim Dey may be the best publicist in the world: She's whip smart and has the amazing ability to make people feel happy and important while she's badgering them. I really don't understand how she does it. To Andrew Wylie, Luke Ingram, and the crew at the Wylie Agency—thank you all for having my back.

While in one sense, the origin of this book may be found in the seminar rooms of Johns Hopkins, in another, it lies in the realm of my own somatic experience. Thanks to Z. for years of transformative loving touch and to my tactile twins, Natalie and Jacob, who started hugging at birth and have, thankfully, never stopped.

NOTES

PROLOGUE

1. J. B. Watson, *The Psychological Care of the Infant and Child* (New York: W. W. Norton & Company, 1924), 81–82. Watson continues: "Try it out. In a week's time you will find how easy it is to be perfectly objective with your child and at the same time kindly. You will be utterly ashamed at the mawk-ish, sentimental way you have been handling it. If you expected a dog to grow up and be useful as a watch dog, a bird dog, a fox hound, useful for anything except a lap dog, you wouldn't dare treat it the way you treat your child."

CHAPTER ONE: THE SKIN IS A SOCIAL ORGAN

1. I wish that today's academic writers were so clear and evocative. From S. E. Asch, "Forming impressions of personality," *Journal of Abnormal and Social Psychology* 41 (1946): 258–90.
2. We call unemotional, potentially dangerous people "cold-blooded." Many of us find it easy to imagine the emotional lives of warm-blooded (homoeo-thermic) mammals like monkeys, dogs, and cats. Yet on some level we imagine that cold fish (or lizards or snakes) are, well, "cold fish." While we know little about the emotional lives of cold-blooded critters (called poiki-lotherms), we should not fall into the simple trap of imagining that theirs must be like ours, or, worse, imagining that they lack emotional lives.
3. This makes sense in a two-pass model of evaluating others. First, you want to distinguish foe from friend (or at least nonfoe), then you want to assess if that individual has the capacity to act on their intentions toward you. For a useful review, see S. T. Fiske, A. J. Cuddy, and P. Glick, "Universal dimen-sions of social cognition: warmth and competence," *Trends in Cognitive Science* 11 (2007): 77–83.
4. R. Feldman, "Maternal touch and the developing infant," in M. J. Herten-stein and S. J. Weiss, eds., *The Handbook of Touch* (New York: Springer, 2011), 373–407.

5. Importantly, just like Asch's 1946 study, the warm coffee cup did not produce a general halo effect of goodwill. Personality descriptors that do not associate with the warm/cold dimension (like reliable/unreliable or honest/dishonest) were unaffected by the temperature of the cup. L. E. Williams and J. A. Bargh, "Experiencing physical warmth promotes interpersonal warmth," *Science* 322 (2008): 606–7.

6. J. M. Ackerman, C. C. Nocera, and J. A. Bargh, "Incidental haptic sensations influence social judgments and decisions," *Science* 328 (2010): 1712–15.

7. It's useful to note that the heavier clipboard will not only produce greater activation of sensors in the skin but will also require a greater degree of contraction in the muscles of the arm, which have their own sensors and pathways to the brain. This issue will be explored in greater detail in later chapters.

8. The experimenters also asked these subjects to rate relationship familiarity of the fictional people in the passage (closeness of relationship, business-style versus casual-style interaction), and these evaluations were unaffected by the prior smooth/rough texture experience.

9. M. W. Kraus, C. Huang, and D. Keltner, "Tactile communication, cooperation and performance: an ethological study of the NBA," *Emotion* 10 (2010): 745–49. It was really fun to read the detailed methods section of this paper. To wit: "Coding focused on intentional forms of touch; thus, contact resulting directly from playing basketball (e.g., fighting for position, setting screens) was not coded. In addition, due to unreliable camera angles, we chose not to code touch during slow-motion playbacks, reaction shots, timeouts, or during the end of game quarters. We focused our analysis on 12 distinct types of touch that occurred when two or more players were in the midst of celebrating a positive play that helped their team (e.g., making a shot). These celebratory touches included fist bumps, high fives, chest bumps, leaping shoulder bumps, chest punches, head slaps, head grabs, low fives, high tens, full hugs, half hugs, and team huddles." It's notable that butt slaps and full-on smooches were not mentioned in this analysis. Two separate coders scored each game, and a touch was scored only if both coders agreed, which occurred for 83 percent of all touches.

10. The performance metrics used in this study are interesting. For example, points scored is not a sophisticated metric: Some players may score a lot of points by taking a lot of shots and making only a moderate percentage of these, thereby turning over the ball. WinScore, the performance measure used here, accounts for the overall positive impact a player has on his team's success by incorporating rebounds, points scored, shooting attempts, assists, blocks, and steals.

11. For a clearly written overview of coalitional behavior and grooming in primates, see Robin Dunbar's fine book *Grooming, Gossip, and the Evolution of Language* (Cambridge, MA: Harvard University Press, 1996), particularly chapter 2. Dunbar sounds the important caveat that not all primate societies exhibit coalitional behavior. Most New World monkeys don't, and this also holds true for some Old World lineages, like colobus monkeys and lemurs. In general, the larger the social group, the more likely that coalitional behavior, reinforced by extensive grooming, will occur.

12. Of course, languorous baboon social life is not all about happy animals grooming one another. They can also be total assholes. Robert Sapolsky,

who studies stress in baboon societies, writes, "Baboons are perfect models for the ecosystem I study. They live in the Serengeti in East Africa, which is a wonderful place for a baboon to live. They're in big troops, so predators don't hassle them much. Infant mortality is low. Most importantly, it takes baboons only about 3 hours of foraging to get their day's calories. Critical implication of this—if you are spending only 3 hours in a day getting food, that means you have 9 hours of free time each day to devote to being miserable to some other baboon. Like us, they are ecologically privileged enough so that they can devote their time to generating psychological stress for each other. *If a baboon in the Serengeti is miserable, it is because another baboon has worked very hard to bring that state about.*" (Italics mine.) From a 2011 interview of Robert Sapolsky by Avi Solomon on the Web site boingboing .net: http://boingboing.net/2011/11/23/robert-sapolsky-on-stress-an.html.

13. G. S. Wilkinson, "Social grooming in the common vampire bat, *Desmodus rotundus*," *Animal Behavior* 34 (1986): 1880–89; G. S. Wilkinson, "Food sharing in vampire bats," *Scientific American* 262 (1990): 76–82.

14. The two flagship papers for this line of work are S. Levine, "Infantile experience and resistance to physiological stress," *Science* 126 (1957): 405; and S. Levine, M. Alpert, and G. W. Lewis, "Infantile experience and the maturation of the pituitary adrenal axis," *Science* 126 (1957): 1347. These first investigations used handling from birth to twenty-one days old, but subsequent studies revealed that the critical window is from birth to seven days old—handling starting after seven days has no significant effect. Could it be that rat pup handling is a form of maternal deprivation? This doesn't seem likely, as over the course of the day mother rats routinely leave their nests for periods of twenty to twenty-five minutes.

15. D. Liu, J. Diorio, B. Tannenbaum, C. Caldji, D. Francis, A. Freedman, S. Sharma, D. Pearson, P. Plotsky, and M. J. Meaney, "Maternal care, hippocampal glucocorticoid receptors, and hypothalamic-pituitary-adrenal responses to stress," *Science* 277 (1997): 1659–62.

16. C. Caldji, B. Tannenbaum, S. Sharma, D. Francis, P. M. Plotsky, and M. J. Meaney, "Maternal care during infancy regulates the development of neural systems mediating the expression of fearfulness in the rat," *Proceedings of the National Academy of Sciences of the United States of America* 95 (1998): 5335–40; and C. Caldji, J. Diorio, and M. J. Meaney, "Variations in maternal care in infancy regulate the development of stress reactivity," *Biological Psychiatry*, 48 (2000): 1164–74.

17. D. Francis, J. Diorio, D. Liu, and M. J. Meaney, "Nongenomic transmission across generations of maternal behavior and stress responses in the rat," *Science* 286 (1999): 1155–58. These experiments had all the clever controls that you would want: pups from low licking-grooming mothers that were cross-fostered to other low licking-grooming mothers and sham-adopted pups that were simply removed from their mothers for a few minutes and then returned.

18. While we don't yet understand all the cellular and molecular steps that intervene between the tactile sensations received by the newborn pup and the lifelong changes to the stress hormone signaling system, some aspects have been revealed. For example, pups with high licking-grooming mothers have a modification of the gene that codes for the glucocorticoid receptor. This

modification, called DNA demethylation, promotes expression of the glucocorticoid receptor gene. The methyl groups attach to a portion of the gene
called the promoter and prevent binding of an accessory protein, the tran-
scription factor NGFI-A. When licking-grooming–triggered demethylation occurs in early postnatal life, this allows for expression of the gene,
ultimately increasing the levels of glucocorticoid receptor protein and
promoting the negative feedback loop that turns down expression of the
hypothalamic stress hormone CRH (see figure 1.5). The key feature of
this mechanism is that it is epigenetic rather than genetic, meaning that
it does not change the series of nucleotides that comprise the individual
rat's genetic code. Rather, it chemically modifies genes to turn their activity up or down. For a nice review of the epigenetics of maternal care in rodents, see A. Kaffman and M. J. Meaney, "Neurodevelopmental sequelae of
postnatal maternal care in rodents: clinical and research implications of molecular insights," *Journal of Child Psychology and Psychiatry* 48 (2007):
224–44.

19. J. K. Rose, S. Sangha, S. Rai, K. R. Norman, and C. H. Rankin, "Decreased
sensory stimulation reduces behavioral responding, retards development,
and alters neuronal connectivity in *Caenorhabditis elegans*," *Journal of
Neuroscience* 25 (2005): 7159–68; S. Rai and C. H. Rankin, "Critical and
sensitive periods for reversing the effects of mechanosensory deprivation on
behavior, nervous system, and development in *Caenorhabditis elegans*,"
Developmental Neurobiology 67 (2007): 1443–56.

20. In affluent households, touch deprivation is likely to occur later in life, with
school-age children. The effects of the later deprivation can be serious but
not as crucial as that occurring in early life.

21. A classic book that discusses the role of touch in development is T. Field,
Touch (Cambridge, MA: MIT Press, 2001). For more recent reviews on this
topic, see M. M. Stack and A. D. Jean, "Communicating through touch:
touching during parent-infant interactions," in *The Handbook of Touch*,
M. J. Hertenstein and S. J. Weiss, eds. (New York: Springer, 2011), 273–98;
and R. Feldman, "Maternal touch and the developing infant," in *The Handbook of Touch*, Hertenstein and Weiss, eds. 373–407.

22. Some recent reviews of kangaroo care include S. Bailey, "Kangaroo mother
care," *British Journal of Hospital Medicine* 73 (2012): 278–81; and J. E.
Lawn, J. Mwansa-Kambafwile, B. L. Horta, F. C. Barros, and S. Cousens,
"Kangaroo mother care to prevent neonatal deaths due to preterm birth
complications," *International Journal of Epidemiology* 39 (2010): 144–54.

23. R. Feldman, Z. Rosenthal, and A. Eidelman, "Maternal-preterm skin-to-skin
contact enhances child physiologic organization and cognitive control across
the first 10 years of life," *Biological Psychiatry* 75 (2014): 56–64.

24. S. E. Jones and A. E. Yarbrough, "A naturalistic study of the meanings of
touch," *Communications Monographs* 52 (1985): 19–56; and F. N. Willis and
C. M. Rinck, "A personal log method for investigating interpersonal touch,"
Journal of Psychology 113 (1983): 119–22.

25. M. J. Hertenstein, D. Keltner, B. App, B. A. Bulleit, and A. R. Jaskolka,
"Touch communicates distinct emotions," *Emotion* 6 (2006): 528–33; M. J.
Hertenstein, R. Holmes, M. McCullough, and D. Keltner, "The communi-
cation of emotion via touch," *Emotion* 9 (2009): 566–73.

26. The world's leader in the detailed analysis of emotional facial expressions is Paul Ekman. A nice review article he wrote is P. Ekman, "Facial expression and emotion," *American Psychologist* 48 (1993): 384–92.

27. M. J. Hertenstein and D. Keltner, "Gender and the communication of emotion via touch," *Sex Roles* 64 (2011): 70–80.

28. For those who are interested in the roles of gender and social status in interpersonal touch, I highly recommend the following review article by Judith Hall, which nicely summarizes and synthesizes a rather difficult and sometimes contradictory literature: J. A. Hall, "Gender and status patterns in social touch," in *The Handbook of Touch*, M. J. Hertenstein and S. J. Weiss, eds. (New York: Springer, 2011), 329–50. Hall concludes that gender differences in touching and decoding touch cannot be simply explained in terms of a model based upon social status.

29. S. M. Jourard, "An exploratory study of body-accessibility," *British Journal of Social and Clinical Psychology* 5 (1966): 221–31. The couples in this study were a mixture of same-sex and opposite-sex pairs. They were not interviewed and they likely represent a mixture of friends and lovers.

30. The British do tend to come out at the bottom of international ranking for social touch. One famous anecdote comes from Prince Charles. He describes a disappointing incident in his boyhood when his mother, the queen, returned from a trip abroad and he was called upon to wait in line with a group of well-wishers for her greeting, which turned out to be only a handshake. Recounted in S. Heller, *The Vital Touch* (New York: Henry Holt and Company, 1997), 155.

31. E. R. McDaniel and P. A. Andersen, "Intercultural variations in tactile communication," *Journal of Nonverbal Communication* 22 (1998): 59–75. The nationalities of the subjects were confirmed by verbal queries after the touches were observed. This study, while interesting, did not have great statistical power and did not sample world cultures broadly: There were no Africans in the data set and few southern Europeans.

CHAPTER TWO: PICK IT UP AND PUT IT IN YOUR POCKET

1. There are several levels of confusion in this question. First, there is a general assumption that sensory experience is a key determinant of intelligence. Yet we know that people who are blind or deaf from birth can have normal intelligence. Second, the precise meanings of phrases like "more acute hearing" and "keener eyesight" are vague. Let's consider some examples. For the frequencies that dominate in human speech, human hearing is extremely sensitive: In this range we can detect faint sounds as well as any other animal. However, there is a range of low-frequency sounds that can be perceived by elephants, whales, rhinos, and prairie chickens that only weakly activates the sensory apparatus in the human middle ear. Likewise, bats, dogs, dolphins, and mice can detect high-pitched sounds that are above the upper range of human hearing (about 20,000 hertz in children). In the case of vision, hawks (and other birds of prey) have both greater visual acuity than humans (in the ophthalmologist's office, a hawk would have about 20/2 vision on an eye chart—it can resolve details at 20 feet that the average human could resolve from only 2 feet away) and a broader range of color vision

(for example, they can see ultraviolet light emanating from the urine trails of their prey that our eyes cannot detect).

2. Aristotle, *De Anima [On the Soul]*, Book 2.

3. Writing in the year 1900, Cornell University psychologist I. M. Bentley waxed rhapsodic about the skin:

> "One of the surprises of physiology is the revelation of the multitude of functions performed by this apparently simple organ. As a rind, it is not only the container but the warder-off, and also the go-between for the organism and its world; tegument, buckler, interagent. It is small wonder that its work is represented in mental process; that many of our most worn and useful perceptions are made up of cutaneous sensations."

> I. M. Bentley, "The synthetic experiment," *American Journal of Psychology* 11 (1900): 405–25.

4. My skin would also be found to contain about 5 million hairs and about 2 million sweat glands. Surprisingly, even for women with a full, thick mane, only about 2 percent of those hairs are found on the scalp, as most are fine vellus hairs. At full throttle, when exercising strenuously in hot and humid conditions, our sweat glands can secrete more than a gallon per hour, making us humans the champion *schvitzers* of the animal kingdom. We may not have the best eyesight or hearing, but we can sweat like crazy. Go *Homo sapiens*! And be sure to stay hydrated.

5. Both the *Oxford English Dictionary* and the *Merriam-Webster* dictionary say that the correct pronunciation of the word *glabrous* is GLAY-bruss. However, not a single dermatologist or skin scientist I have met uses this pronunciation. They all say GLAB-russ, except for some of the British scientists who say GLAH-bruss. Given this confusion, I think that you can comfortably use any one of these three pronunciations.

6. Some doctors use a special term for that subset of glabrous skin that borders mucous membranes. In this terminology, mucocutaneous skin includes the lips, conjunctivae (external membranes of the eyes), labia minora, skin surrounding the urethra and anus, etc. While the nipple is entirely glabrous, only the central part of the areola is glabrous, with hairy skin at the periphery. Junot Diaz riffs about this in his splendid award-winning novel *The Brief Wondrous Life of Oscar Wao*: "Your mother's breasts are immensities... They're 35 triple-D's and the aureoles are as big as saucers and black as pitch and at their edges are fierce hairs that sometimes she plucked and sometimes she didn't."

7. For example, the epidermis of the glabrous skin of the fingertips is about ten times thicker than that of the hairy skin of the forearm. The epidermis is thickest on the glabrous skin of the sole, about 1 millimeter, which is about twenty-fold thicker than that of the forearm. This is because the skin of the sole must endure great mechanical stress during walking and running. It's not surprising that the time it takes for new cells to completely renew the epidermis is longer in those locations where it is very thick. The dermis also varies in thickness over the body surface, ranging from about 1 millimeter in the armpit to about 2.5 millimeters on the back.

8. J. K. McNeely, *Holy Wind in Navajo Philosophy* (Tucson: University of Arizona Press, 1981), 35.

9. M. J. Henneberg, K. M. Lambert, and C. M. Leigh, "Fingerprint homoplasy: koalas and humans," *NaturalSCIENCE.com* 1, article 4 (1997), found at http://naturalscience.com/ns/articles/01-04/ns_hll.html. These authors also mention that fingerprint-like patterns are found on the prehensile tails of some mammalian species. Also see M. Okajima, "Nonprimate mammalian dermatoglyphics as models for genetic and embryological studies: comparative and methodologic aspects," *Birth Defects Original Articles Series* 27 (1991): 131–49.

10. T. Lewis and G. W. Pickering, "Circulatory changes in the fingers in some diseases of the nervous system, with special reference to digital atrophy of peripheral nerve lesions," *Clinical Science* 2 (1936): 149.

11. Interestingly, glabrous skin that does not have sympathetically controlled sweat glands, like the penile and clitoral glans, does not wrinkle after water immersion.

12. M. Changizi, R. Weber, R. Kotecha, and J. Palazzo, "Are wet-induced wrinkled fingers primate rain treads?" *Brain, Behavior and Evolution* 77 (2011): 286–90. Their hypothesis is supported by the observation that the particular patterns of ridges and valleys in wet-wrinkled fingers and toes function well for water drainage and mimic the arrangement of natural drainage features found on rain-carved mountainsides, albeit on a much smaller scale.

13. K. Kareklas, D. Nettle, and T. V. Smulders, "Water-induced finger wrinkles improve handling of wet objects," *Biology Letters* 9 (2013): 20120999. However, when another group of authors set out to replicate the experiments of Kareklas et al. (2013), they were unable to do so: J. Haseleu, D. Omerbasic, H. Frenzel, M. Gross, and G. R. Lewin, "Water-induced finger wrinkles do not affect touch acuity or dexterity in handling wet objects," *PLOS One* 9 (2014): e84949.

14. Electrical spikes (also called action potentials) are the main mode of long-distance information transfer in almost all neurons, not just those that convey information from the skin to the spinal cord and brain. At rest, most neurons have a voltage difference across their outer membranes of about -70 millivolts. When a neuron is depolarized to a level of about -55 millivolts, voltage-sensitive ion channels open, letting sodium ions rush in. Because sodium ions are positively charged, their entry into the neuron depolarizes it further, leading to more sodium-channel opening, in a positive feedback loop, causing the rapid upstroke of the spike. About a millisecond later, voltage-sensitive potassium channels open and the sodium channels close, causing potassium to rush out, thereby underlying the downstroke of the spike. Importantly, spikes can propagate from one patch of membrane to another, like a flame traveling along a fuse, igniting the next section as it moves along. This is how spikes travel along nerve fibers from the skin to the spinal cord, and ultimately up to the brain. If you'd like a somewhat more complete explanation of spike signaling in neurons, together with some diagrams, see this section of my earlier book: D. J. Linden, *The Accidental Mind* (Harvard/Belknap Press, 2007), 28–49.

15. It should be noted that there are several different families of stretch-activated ion channels and that they are present in many cell types ranging from white blood cells to kidneys. In the nervous system, stretch-activated channels are also important in the hair cells of the inner ear, where they

help to convert the mechanical energy of sound waves into electrical signals that travel to the brain. The molecular identity of the stretch-activated ion channels that underlie touch perception is not completely understood. At present, the best candidates are proteins called Piezo1 and Piezo2. B. Coste, B. Xiao, J. S. Santos, R. Syeda, J. Grandl, K. S. Spencer, S. E. Kim, M. Schmidt, J. Mathur, A. E. Dubin, M. Montal, and A. Patapoutian, "Piezo proteins are pore-forming subunits of mechanically activated channels," *Nature* 483 (2012): 176–81. Good summaries of the current state of the struggle are found here: D. M. Bautista and E. A. Lumpkin, "Probing mammalian touch transduction," *Journal of General Physiology* 138 (2011): 291–301; and B. Nilius and E. Honoré, "Sensing pressure with ion channels," *Trends in Neuroscience* 35 (2012): 477–86.

16. The original description by Merkel is F. S. Merkel, "Tastzellen und Tastkörperchen bei den Hausthieren und beim Menschen," *Archiv für mikroskopische Anatomie* 11 (1875): 636–52. The title translates as "Touch cells and corpuscles in domestic animals and humans." While Merkel named his newly discovered cells *tastzellen*, arguments about whether they really function as touch sensors went on for years; 134 years, to be exact. In 2009, Huda Zogbhi and collaborators finally put this issue to rest by using genetic engineering to create mice that lacked Merkel cells. Recordings from nerve fibers in these mice showed a lack of characteristic sustained light-touch responses. Sometimes science requires a lot of patience. S. M. Maricich, S. A. Wellnitz, A. M. Nelson, D. R. Lesniak, G. J. Gerling, E. A. Lumpkin, and H. Y. Zogbhi, "Merkel cells are essential for light-touch responses," *Science* 324 (2009): 1580–82. If we're to dig a bit deeper, there are still fundamental questions that remain. Where is the precise location where the force of skin indentation is transduced into an electrical signal? Is it in the Merkel cells, in the Merkel cell-contacting nerve fibers, or both? When Merkel cells are genetically deleted, this abolishes electrical responses in the touch-sensing nerve fiber. Yet there are several possible explanations for this result:

   (a) Merkel cells are required in a mechanical role to transmit force properly to the nerve fiber, but it is the nerve fiber that transduces force into electrical spiking through Piezo channels. Therefore, when you remove Merkel cells, there's no response in the nerve fiber because the ending of the fiber is not properly activated mechanically.

   (b) Merkel cells transduce force into electrical signals, and then they release a chemical signal (a neurotransmitter) to evoke an electrical signal in the nerve fiber.

   (c) When Merkel cells are removed in the mutant mouse by genetic engineering, there is a developmental side effect that makes the nerve fiber unable to transduce force into electrical signals, even though it does so in normal mice (and presumably humans, too).

While the manuscript for this book was in the editing stage, a new report appeared that shed light on this interesting problem. Genetic engineering was used to create a mouse in which the stretch-activated Piezo2 ion channel was deleted in skin cells (including Merkel cells) but not in sensory nerves. In these mice, the Merkel cells no longer had touch-sensitive electrical currents. However, the sensitivity of the glabrous skin to fine mechanical stimulation was decreased but not abolished in these mice. This suggests a two-site model for mechanotransduction in both the Merkel cells and the sensory neurons that contact them, a sort of hybrid of models (a) and (b) above. S. H. Woo, S. Ranade, A. D. Weyer, A. E. Dubin, Y. Baba, Z. Qiu, M. Petrus, T. Miyamoto, K. Reddy, E. A. Lumpkin, C. L. Stucky, and A. Patapoutian, "Piezo2 is required for Merkel-cell mechanotransduction," *Nature* 50 (2014): 622-26.

17. Å. B. Vallbo, K. A. Olsson, K.-G. Westberg, and F. J. Clark, "Microstimulation of single tactile afferents from the human hand," *Brain* 107 (1984): 727–49. This technique requires amazing dexterity on the part of the experimenter and great tolerance on the part of the subject. It involves manually inserting a very fine electrode (with a tip size of 0.01 millimeter) into the arm to carefully seek out a single tactile nerve fiber originating from the hand for single fiber recording and, in some cases, single fiber stimulation. These experiments can take hours to complete. Remarkably, stimulation of single mechanosensory nerve fibers can reliably give rise to both a clear perception of touch and a clear activation of certain touch-processing regions of the brain, as revealed by brain imaging or EEG recording. M. Trulsson and G. K. Essick, "Sensations evoked by microstimulation of single mechanoreceptive afferents innervating the human face and mouth," *Journal of Neurophysiology* 103 (2010): 1741–47; and M. Trulsson, S. T. Francis, E. F. Kelly, G. Westling, R. Bowtell, and F. McGlone, "Cortical responses to single mechanoreceptive afferent microstimulation revealed with fMRI," *NeuroImage* 13 (2001): 613–22.

18. While each Merkel ending is innervated by a single sensory axon, a single axon will typically innervate 10–50 Merkel disks spread over a diameter of 1 to 3 millimeters. It may seem trivial that Merkels can respond weakly to small indentations and strongly to large ones (and do so linearly), but this turns out to be crucial for sensing the curvature of objects—in distinguishing the smooth edge of a penny from that of a nickel, for example. If we were to zoom in on your fingertip contacting the edge of a penny, what we'd see is that the skin of the fingertip is indented most strongly at the central point of contact and the degree of indentation falls off with distance as we move away from that point in both directions. Of course, the rate at which the indentation falls off with distance will be proportional to the curvature of the object—smaller for the nickel and larger for the penny. Because the dense array of Merkel disk endings can accurately report the degree of indentation at each point, sufficient information can be conveyed by a population of Merkel disk nerve fibers to allow the brain to estimate object curvature.

19. If you're starting to suspect that German anatomists of the nineteenth century had an outsize role in describing the cellular structure of organs, you're right. The Meissner's corpuscle was codiscovered by Georg Meissner and his mentor Rudolf Wagner of Göttingen University, and described in an 1852 publication. The next year, Meissner published on these structures again, this time leaving Wagner's name off the work. A bitter fight over priority ensued and the conflict remained unresolved at the time of Wagner's death in 1864.

20. Corpuscle, from the Latin *corpusculum*, or "tiny body," is an odd and vague word. In biology, it can mean either a single free-floating cell, like a red or white blood cell, or a small collection of cells into a particular structure, as in the Meissner's corpuscle. In a humorous intersection of biology, religion, and academe, members of Corpus Christi College, Oxford, or Corpus Christi College, Cambridge, are also called corpuscles. This is the same type of linguistic collision that leads to a delightful official title for the Roman Catholic pontiff: the Supreme Primate.

21. In an inspiration to future doctors everywhere, the Pacinian corpuscle was discovered by Italian anatomist Filippo Pacini when he was still a medical student at Pistoia (a small city in Tuscany) in 1831 doing his assigned cadaver dissections. He first published his findings some years later: F. Pacini, *Nuovi organi scoperti nel corpo umano*, Tipografia Cino, Pistoia, Italy, 1840. The Pacinian corpuscle is also sometimes called the lamellar corpuscle, in reference to the many sheets, or lamellae, of cells that form its capsule.

22. Their concentric lamellar structure also calls to mind that great Edward Weston photograph of an artichoke in cross-section: http://www.masters-of-photography.com/W/weston/weston_artichoke_halved_full.html; or certain Georgia O'Keeffe paintings, like *Grey Line with Black, Blue, and Yellow* (1923): http://www.wikipaintings.org/en/georgia-o-keeffe/gray-line-with-black-blue-and-yellow.

23. Of course, earthquakes and bomb tests have characteristic frequency information as well, so in order to simulate them, jumping children would need an unusual degree of motor control. It's a thought experiment, really.

24. L. Ulrich, "Porsche's baby turns 16; seeks a bigger allowance," *New York Times*, August 17, 2012, http://www.nytimes.com/2012/08/19/automobiles/autoreviews/porsches-baby-turns-16-seeks-a-bigger-allowance.html?pagewanted=all.

25. The Ruffini ending was first described by the Italian anatomist Angelo Ruffini: A. Ruffini, "Les expansions nerveuses de la peau l'homme et quelques autres Mammifeves," *Review of General Histology* 3 (1905): 421–540.

26. The presence of Ruffini endings in the glabrous skin of the hand is still a matter of some debate. For example, they don't appear to be present in the finger pads of the monkey, only at the base of the fingernails. M. Pare, A. M. Smith, and F. L. Rice, "Distribution and terminal arborizations of cutaneous mechanoreceptors in the glabrous finger pads of the monkey," *Journal of Comparative Neurology* 445 (2002): 347–59.

27. While the nerve fibers that convey information from the mechanosensors do not blend signals from more than one type of detector, these "labeled lines" are not perfectly maintained all the way to the brain. At every relay station in the spinal cord and the brain itself, there is some degree of signal blending. As we will see in later chapters, this signal blending can also cross touch modalities: There are neurons in the spinal cord that are driven by both light touch (mechanosensation) and temperature or light touch and pain, for example.

28. Z. gets miffed if I space out and caress her arm against the grain of the hairs. "If you did that to a cat it would bite you, or at least walk away," she says.

29. V. E. Abraira and D. D. Ginty, "The sensory neurons of touch," *Neuron* 79 (2013): 618–39.

30. For comparison, average sighted reading speed in English is about 250 words per minute. In both sighted and Braille reading there is a trade-off between speed and accurate comprehension.

31. J. R. Phillips, R. S. Johansson, and K. O. Johnson, "Representation of braille characters in human nerve fibers," *Experimental Brain Research* 81 (1990): 589–92; F. Vega-Bermudez, K. O. Johnson, and S. S. Hsiao, "Human tactile pattern recognition: active versus passive touch, velocity effects, and patterns of confusion," *Journal of Neurophysiology* 65 (1991): 531–46.

32. I know, I know—it's kinda kinky. But we all must sacrifice for the truth. And because I know that you're wondering, I'll make it clear: She did not try doing two-point threshold discrimination on her clitoris.

33. The two-point discrimination threshold was a standard neurological test for many years, but it tends to systematically underestimate tactile resolution. In most labs it has since been replaced by a task in which gratings bearing raised ridges are pressed into the skin with constant force. The distance between the ridges is systemically varied and the subject is asked whether the ridges are running horizontally or vertically. K. O. Johnson and J. R. Phillips, "Tactile spatial resolution I. Two-point discrimination, gap detection, grating resolution and letter recognition," *Journal of Neurophysiology* 46 (1981): 1177–92; R. W. Van Boven and K. O. Johnson, "The limit of tactile spatial resolution in humans: grating orientation discrimination at the lips, tongue and finger," *Neurology* 44 (1994): 2361–66.

34. In the case of the neck and the head, the dorsal root ganglia are replaced by similar structures called the trigeminal ganglia. The technical term for neurons that have this shape is "pseudo-unipolar neurons."

35. This assumes that the mechanosensory axons from our giant conduct signals at the same speed as those of conventional people.

36. For you hard-core anatomy mavens: Neurons that carry information from the mechanoreceptors have axons that ascend in the region of the spinal cord called lamina IV of the dorsal horn. Mechanoreceptor axons from the lower body, below the seventh thoracic vertebra, contact neurons in the gracile nucleus of the brain stem, while those of the upper body form synapses on neurons in the adjacent cuneate nucleus. The gracile and cuneate neurons send their axons to a particular subdivision of the thalamus called the ventroposterolateral region through a midline-crossing pathway called the medial lemniscus. These thalamic cells then project to the primary somatosensory cortex. In later chapters, we'll discuss skin sensors for erotic touch, pain, itch, and temperature, which take a different path in both the spinal cord and the brain.

    Also, a note on terminology: A bundle of axons is called a nerve when it's running through the body (also called the periphery), but once it has entered the spinal cord or the brain, it's called a tract.

37. Crossing of fibers in sensory and motor systems is widespread in the nervous system. In some cases, like the visual system, it makes sense: Your right visual field will activate the left side of the retina on both of your eyes. Therefore, in order to bring information from the right and left halves of the visual field together in the visual cortex, fibers from parts of the retina near the midline must cross in the brain. In the touch system, the reason for crossing is less clear and is sometimes a topic of argument when neuroscientists drink beer.

38. In Penfield's day there weren't effective antiepileptic drugs and these surgeries were performed only when the seizures were so severe that they were completely life-disruptive and dangerous. Penfield's original description of brain mapping is still useful today: W. Penfield and T. C. Erickson, *Epilepsy and Cerebral Localization: A Study of the Mechanism, Treatment and Prevention of Epileptic Seizures* (Springfield, IL: Charles C. Thomas, 1941). For a clear and evocative description of Penfield's surgeries as well as a broader meditation on the function of sensory and motor maps in the brain, see: S. Blakeslee and M. Blakeslee, *The Body Has a Mind of Its Own* (New York: Random House, 2008).

39. More recently, the primary somatosensory touch map has been explored with brain-scanning machines that can detect local brain activity in response to touch. It has also been revealed in laboratory animals by recording the electrical activity of individual neurons in the somatosensory cortex and then determining the location on the skin surface (on the opposite side of the body) where a touch will activate them. It turns out that maps are a ubiquitous feature of sensory systems in the brain and, like the touch map, are often distorted. The primary visual cortex has a map of the visual world (that's upside down and backward) in which the center of the retina, which is most densely packed with light-detecting cells, is magnified. The auditory cortex has a map of pitch. Maps for the chemical senses, like smell, have been harder to understand, as odor-space is not as easily grasped as skin-space or retina-space.

40. Fans of the horror writer H. P. Lovecraft have been known to compare star-nosed moles to Lovecraft's extraterrestrial creature Cthulhu, first described in a 1928 story: "A monster of vaguely anthropoid outline, but with an octopus-like head whose face was a mass of feelers, a scaly, rubbery-looking body, prodigious claws on hind and fore feet, and long, narrow wings behind." Okay, so the mole doesn't have wings, but still.

41. The world leader in studying the star-nosed mole is Kenneth Catania of Vanderbilt University. He has used high-speed video to show that when foraging, star-nosed moles search with their rays in a random fashion. But once a ray detects a potential food source, the star twitches so that a specialized supersensitive ray, number 11, now contacts the object. One or more touches with ray number 11 are typically required to determine whether the object really is prey and therefore should be consumed. When Professor Catania carefully examined the star-nosed mole's cortical touch map, he found that not only did the star occupy an outsize portion of the somatosensory cortex but that within the star, number 11 occupied the greatest space of any ray. A nice review of this work is K. C. Catania, "The sense of touch in the star-nosed mole: from mechanoreceptors to the brain," *Philosophical Transactions of the Royal Society of London, Series B* 366 (2011): 3016–25.

42. Cool nerdly detail: If you look at the skin in the finger pad you'll see that a single sensory axon carries information from an array of Merkel endings in the skin spread over 2 to 3 millimeters in diameter. However, if you make an electrical recording from this axon while probing the finger pad with a fine-pointed instrument, you'll find that it tends to respond to stimulation in a smaller area, about 1 millimeter in diameter. This is called the receptive field of the axon. The disparity between the receptive field and the

anatomical distribution of endings suggests that something interesting is going on. Perhaps the Merkels at the edge of the array signal only weakly or perhaps their signals are actively blocked as the branches of the axon converge coursing toward the spinal cord.

43. The discovery of columnar organization of the neocortex was pioneered by Vernon Mountcastle and his coworkers at the Johns Hopkins University School of Medicine. A nice summary of this work may be found here: V. B. Mountcastle, *The Sensory Hand* (Cambridge, MA: Harvard University Press, 2005), 260–300.

44. The organization of the secondary somatosensory cortex is poorly understood, but it is becoming clear that, like the primary somatosensory cortex, it is composed of functionally distinct areas, some of which receive only tactile (cutaneous) signals and others that receive a mixture of tactile signals and information about the conformation of the hand/limb/body (proprioceptive signals). P. J. Fitzgerald, J. W. Lane, P. H. Thakur, and S. S. Hsiao, "Receptive field properties of the macaque second somatosensory cortex: evidence for multiple functional representations," *Journal of Neuroscience* 24 (2004): 11193–204.

45. This was clearly shown in a series of experiments by Esther Gardner and her colleagues at New York University School of Medicine, who recorded from the posterior parietal cortex in monkeys trained to reach and grasp objects. They found that neurons in this region became active while the monkey was reaching for an object, long before the object was touched. Furthermore, the firing patterns of these neurons during reaching appeared to reflect predictions about the anticipated efficiency of the reaching movements. The neural responses after contacting an object can then either confirm or rebut these predictions, thereby providing the information for the monkey to learn to optimize its reaching movements as a result of experience, i.e., motor learning. J. Chen, S. D. Reitzen, J. B. Kohlenstein, and E. P. Gardner, "Neural representation of hand kinematics during prehension in posterior parietal cortex of the macaque monkey," *Journal of Neurophysiology* 102 (2009): 3310–28.

46. T. Elbert, C. Pantev, C. Weinbruch, B. Rockstroh, and E. Taub, "Increased cortical representation of the fingers of the left hand in string players," *Science* 270 (1995): 305–7; I. Hashimoto, A. Suzuki, T. Kimurs, Y. Iguchi, M. Tanosaki, R. Takino, Y. Haruta, and M. Taira, "Is there training-dependent reorganization of digit representations in area 3b of string players?" *Clinical Neurophysiology* 115 (2004): 435–47; and P. Schwenkreis, S. El Tom, P. Ragert, B. Pleger, M. Tegenthoff, and H. R. Dinse, "Assessment of sensorimotor cortical representation asymmetries and motor skills in violin players," *European Journal of Neuroscience* 26 (2007): 3291–3302. What are the functional consequences of having an expanded left hand representation in the touch map? The last study went on to show that motor performance was not enhanced in either the left or right hand of musicians as compared with nonmusicians. These authors did not test tactile discrimination, however.

47. C. Xerri, J. M. Stern, and M. M. Merzenich, "Alterations of the cortical representation of the rat ventrum induced by nursing behavior," *Journal of Neuroscience* 14 (1994): 1710–21; and C. Rosselet, Y. Zennou-Azogui, and C. Xerri, "Nursing-induced somatosensory plasticity: temporally decoupled

changes in neuronal receptive field properties are accompanied by modifications in activity-dependent protein expression," *Journal of Neuroscience* 26 (2006): 10667–76. Inquiring minds want to know: Does this map plasticity also occur in breast-feeding humans?

48. J. O. Cog and C. Xerri, "Tactile impoverishment and sensorimotor restriction deteriorate the forepaw cutaneous map in the primary somatosensory cortex of adult rats," *Experimental Brain Research* 129 (1999): 518–31.

49. C. F. Bolton, R. K. Winkelmann, and P. J. Dyck, "A quantitative study of Meissner's corpuscles in man," *Neurology* 16 (1966): 1–9; M. F. Bruce, "The relation of tactile thresholds to histology in the fingers of elderly people," *Journal of Neurology, Neurosurgery and Psychiatry* 43 (1980): 730–34; J. C. Stevens and K. K. Choo, "Spatial acuity of the body surface over the life span," *Somatosensory and Motor Research* 13 (1996): 153–66; and K. L. Woodward, "The relationship between skin compliance, age, gender and tactile discriminative thresholds in humans," *Somatosensory and Motor Research* 10 (1993): 63–67

50. R. W. Van Boven, R. H. Hamilton, T. Kauffman, J. P. Keenan, and A. Pascual-Leone, "Tactile spatial resolution in blind braille readers," *Neurology* 54 (2000): 2230-2236; and D. Goldreich and I. M. Kanics, "Tactile acuity is enhanced in blindness," *Journal of Neuroscience* 23 (2003): 3439–45. The observation of superior performance by women in these tactile acuity tasks was an outgrowth of the main topic of these studies, effects of blindness. We'll address the effects of blindness on touch perception in later chapters.

51. R. M. Peters, E. Hackeman, and D. Goldreich, "Diminutive digits discern delicate details: fingertip size and the sex difference in tactile spatial acuity," *Journal of Neuroscience* 29 (2009): 15756–61; and M. Wong, R. M. Peters, and D. Goldreich, "A physical constraint on perceptual learning: tactile spatial acuity improves with training to a limit set by finger size," *Journal of Neuroscience* 33 (2013): 9345–52. Of course, there are some caveats. First and foremost, the assumption that sweat pore density is a good proxy for Merkel disk density awaits confirmation. Second, while, statistically speaking, fingertip size could explain most of the variation in tactile acuity, this finding does not rule out the possibility that there are also sex differences in the brain's touch circuits that also contribute to tactile acuity differences. Goldreich and colleagues went on to ask, on average, do children, with their tiny fingers, have even better tactile acuity than women? The answer is interesting. On average, children's sweat pores, and hence, presumably, their Merkel disks, are indeed spaced more closely. And at a given age, children with larger fingertips had significantly poorer acuity. However, acuity did not worsen with age in children as their fingertips grew. The authors propose that while finger growth in childhood reduces density of the Merkels, this is compensated for by a simultaneous increase in the ability of the brain to extract information from these sensors. R. M. Peters and D. Goldreich, "Tactile spatial acuity in childhood: effects of age and fingertip size," *PLOS One* 8 (2013): e84650.

52. Are there a fixed number of mechanosensors on the female breast? In my limited erotic experience, women with smaller breasts also tend to be more responsive to gentle stimulation of that area. When I shared this idea with Z., she observed, "The size of the breast isn't relevant because the tactile hot spot is the nipple and areola. The surrounding skin of the breast is much less

sensitive." She pointed out that while nipple size and breast size aren't correlated, areola size and breast size are. So it's possible that women with larger breasts have decreased mechanosensor density in the areola. However, it's also possible that mechanosensor density has little to do with erotic response. We'll talk about that in the next two chapters. Later, I came across the following publications. They report that many women seeking breast reduction surgery complain of a lack of sensation in the nipple/areola region. These anecdotal reports were confirmed experimentally: When tactile sensitivity of the areola was measured using standard filaments, two-point discrimination thresholds or vibration, large-breasted women (D-cup size or larger) were consistently less sensitive than small-breasted women (A or B cup). S. Slezak and A. L. Dellon, "Quantitation of sensibility in gigantomastia and alteration following reduction mammoplasty," *Plastic and Reconstructive Surgery* 91 (1993): 1265–69; and Y. Godwin, K. Valassiadou, S. Lewis, and H. Denley, "Investigation into the possible cause of subjective decreased sensory perception in the nipple-areola complex of women with macromastia," *Plastic and Reconstructive Surgery* 113 (2004): 1598–1606.

CHAPTER THREE: THE ANATOMY OF A CARESS

1. If this were a proper Tarantino revenge fantasy, we'd be doing a biopsy of the pudendal nerve, the one that innervates the penis (without anesthetic, I might add). However, that one is not a pure sensory nerve, as it has motor and sympathetic axons running in it. So I've chosen the sural nerve to make the explanation more straightforward. Pedagogy trumps poetry, I guess.

2. A-delta fibers, being lightly myelin-wrapped and intermediate in diameter, have a range of conduction velocities in between those of A-beta fibers and C-fibers: around 10 to 70 miles per hour.

3. The C-tactile neurons and their associated axonal fibers identified in this paper are molecularly distinct from other sensory neurons because they express an enzyme called tyrosine hydroxylase. L. Li, M. Rutlin, V. E. Abraira, C. Cassidy, L. Kus, S. Gong, M. P. Jankowski, W. Luo, N. Heintz, H. R. Koerber, C. J. Woodbury, and D. D. Ginty, "The functional organization of cutaneous low-threshold mechanosensory neurons," *Cell* 147 (2011): 1615–27. It is possible that there are other populations of C-tactile neurons. A different group has shown that a particular receptor called MrgprB4 defines a unique population of C-fibers that innervate hairy but not glabrous skin and thereby are potential candidates for C-tactile fibers. Unfortunately, to date, recordings from these fibers with electrodes have failed to reveal electrical activity evoked by light stroking of the hairy skin, so the role of MrgprB4-expressing C-fibers in caress detection remains unclear. Q. Liu, S. Vrontou, F. L. Rice, M. J. Zylka, X. Dong, and D. J. Anderson, "Molecular genetic visualization of a rare subset of unmyelinated sensory neurons that may detect gentle touch," *Nature Neuroscience* 10 (2007): 946–48; and S. Vrontou, A. M. Wong, K. K. Rau, H. R. Koerber, and D. J. Anderson, "Genetic identification of C-fibers that detect massage-like stroking of hairy skin in vivo," *Nature* 493 (2013): 669–73.

4. A. B. Sterman, H. H. Schaumburg, and A. K. Asbury, "The acute sensory neuronopathy syndrome: a distinct clinical entity," *Annals of Neurology* 7

(1980): 354–58. Acute sensory neuronopathy is not heritable and is permanent. It is not typically associated with disruption of immune function. In many, but not all, cases, it has followed treatment with high doses of certain antibiotics such as penicillin or related compounds.

5. O. Sacks, "The Disembodied Lady," in *The Man Who Mistook His Wife for a Hat* (London: Duckworth, 1985), 42–52.

6. H. Olausson, J. Cole, K. Rylander, F. McGlone, Y. Lamarre, B. G. Wallin, H. Kramer, J. Wessberg, M. Elam, M. C. Bushnell, and Å. Vallbo, "Functional role of unmyelinated tactile afferents in human hairy skin: sympathetic response and perceptual localization," *Experimental Brain Research* 184 (2008): 135–40.

7. L. S. Löken, J. Wessberg, I. Morrison, F. McGlone, and H. Olausson, "Coding of pleasant touch by unmyelinated afferents in humans," *Nature Neuroscience* 12 (2009): 547–48; and R. Ackerley, E. Eriksson, and J. Wessberg, "Ultra-late EEG potential evoked by preferential activation of unmyelinated tactile afferents in human hairy skin," *Neuroscience Letters* 535 (2013): 62–66.

8. H. Olausson, Y. Lamarre, H. Backlund, C. Morin, B. G. Wallin, G. Starck, S. Ekholm, I. Strigo, K. Worsley, Å. B. Vallbo, and M. C. Bushnell, "Unmyelinated tactile afferents signal touch and project to insular cortex," *Nature Neuroscience* 5 (2002): 900–904; M. Björnsdottir, L. Löken, H. Olausson, Å. Vallbo, and J. Wessberg, "Somatotopic organization of gentle touch processing in the posterior insular cortex," *Journal of Neuroscience* 29 (2009): 9314–20; and I. Morrison, M. Björnsdottir, and H. Olausson, "Vicarious responses to social touch in posterior insular cortex are tuned to pleasant caressing speeds," *Journal of Neuroscience* 31 (2011): 9554–62.

9. There's some argument about the historical details here. Some people in Norrbotten believe that the pain-insensitive founder lived even earlier, perhaps before the ancestors of the affected Norrbotten families moved to Norrbotten from southern Finland in the sixteenth century.

10. The Norrbotten pain-insensitivity syndrome is called hereditary sensory and autonomic neuropathy type V, abbreviated as HSAN V. E. Einarsdottir, A. Carlsson, J. Minde, G. Toolanen, O. Svensson, G. Solders, G. Holmgren, D. Holmberg, and M. Holmberg, "A mutation in the nerve growth factor beta gene (*NGFB*) causes loss of pain perception," *Human Molecular Genetics* 13 (2004): 799–805; J. Minde, G. Toolanen, T. Andersson, I. Nennesmo, I. N. Remahl, O. Svensson, and G. Solders, "Familial insensitivity to pain (HSAN V) and mutation in the *NGFB* gene. A neurophysiological and pathological study," *Muscle & Nerve* 30 (2004): 752–60; and D. C. de Andrade, S. Baudic, N. Attal, C. L. Rodrigues, P. Caramelli, A. M. M. Lino, P. E. Marchiori, M. Okada, M. Scaff, D. Bouhassira, and M. J. Teixeira, "Beyond neuropathy in hereditary sensory and autonomic neuropathy type V: cognitive evaluation," *European Journal of Neurology* 15 (2008): 712–19.

11. I. Morrison, L. S. Löken, J. Minde, J. Wessberg, I. Perini, I. Nessesmo, and H. Olausson, "Reduced C-afferent fibre density affects perceived pleasantness and empathy for touch," *Brain* 134 (2011): 1116–26. Interestingly, there may also be a rodent model for loss of C-tactile sensation: Naked mole rats comprise about 20 species that live in Africa and are almost entirely devoid of hairs. This all-glabrous condition results in them having only about 25 percent of the number of C-fibers when compared with other mole rats that

bear fur. Presumably the surviving C-fibers in naked mole rats are for slow pain and temperature sensation. E. St. John Smith, B. Purfurst, T. Grigoryan, T. J. Park, N. C. Bennett, and G. R. Lewin, "Specific paucity of unmyelinated C-fibers in cutaneous peripheral nerves of the African naked mole rat: comparative analysis using six species of Bathyergidae," *Journal of Comparative Neurology* 520 (2012): 2785–2803.

12. There are some important caveats about the role of the C-tactile system in social touch. Most of the experiments in support of a special role for the C-tactile system in social touch are done in humans, where we have limited ability to record neural signals and to trace the anatomy of C-tactile fibers. For example, in mice, we know that C-tactile fibers form lanceolate endings on hair follicles and are activated by touches that bend hairs. We don't yet know if this is also the case in humans. It is likely that the C-tactile nerve fibers also convey touch signals that are unrelated to social touch. It is also likely that there are aspects of social touch that are not conveyed by the C-tactile system. We know that C-tactile fibers do not innervate glabrous skin, yet touches on glabrous skin can have an important social role for humans: just think of hand-holding or handshaking.

13. There are two nice review articles about the C-tactile system from the same group in Sweden: I. Morrison, L. S. Löken, and H. Olausson, "The skin as a social organ," *Experimental Brain Research* 204 (2010): 305–14; and M. Björnsdottir, I. Morrison, and H. Olausson, "Feeling good: on the role of C fiber mediated touch in interoception," *Experimental Brain Research* 207 (2010): 149–55.

14. In this study, the subjects were all heterosexual males who were led to believe that they were being sensually caressed by either a man or a woman. In fact, the caresses were all delivered by a woman. V. Gazzola, M. L. Spezio, J. A. Etzel, F. Castelli, R. Adolphs, and C. Keysers, "Primary somatosensory cortex discriminates affective significance in social touch," *Proceedings of the National Academy of Sciences of the USA* (2012): E1657-E1666.

15. I. Gordon, A. C. Voos, R. H. Bennett, D. Z. Bolling, K. A. Pelphrey, and M. D. Kaiser, "Brain mechanisms for processing affective touch," *Human Brain Mapping* (2011), doi:10.1002/hbm.21480; F. McGlone, H. Olausson, J. A. Boyle, M. Jones-Gotman, C. Dancer, S. Guest, and G. Essick, "Touching and feeling: differences in pleasant touch processing between glabrous and hairy skin in humans," *European Journal of Neuroscience* 35 (2012): 1782–88; and L. Lindgren, G. Westling, C. Brulin, S. Lehtipalo, M. Andersson, and L. Nyberg, "Pleasant human touch is represented in pregenual anterior cingulate cortex," *NeuroImage* 59 (2012): 3427–32.

16. A. C. Voos, K. A. Pelphrey, and M. D. Kaiser, "Autistic traits are associated with diminished neural response to affective touch," *Social Cognitive and Affective Neuroscience,* advance electronic access, 2012.

17. These experiments are reported in Morrison et al. (2011a and 2011b), as cited earlier in this chapter (notes 8 and 11).

CHAPTER FOUR: SEXUAL TOUCH

1. Some of our expectations about the world are genetically determined and experience-independent. For example, we are hardwired to expect light

sources to exist in the top half but not the bottom half of our visual field. Hence the old tent-camping trick of making a ghoulish face by shining a flashlight upward from your lap.

2. Of course, there are some who do find the touch of a doctor's exam arousing. This variation in human behavior actually reinforces the main point: The ultimate perception of a stimulus is molded by an individual's prior experience. It's just that people have different life experiences and are shaped by them in different ways.

3. The free nerve endings derive from slowly conducting A-delta and C-fibers. We'll explore the function of these pain and temperature-transducing fibers in detail in chapters 5 and 6.

4. While these structures had been known since at least 1866, the term "genital end bulb" was first used in this paper: A. S. Dogiel, "Die Nervenendigungen in det Haut der aussern Genitalorgane des Menschen," *Archiv fur Mikroscopische Anatomie Forschung* 41 (1893): 585–612.

5. R. K. Winkelmann, "The erogenous zones: their nerve supply and significance," *Proceedings of the Staff Meetings of the Mayo Clinic* 34 (1959): 39–47; K. E. Krantz, "Innervation of the human vulva and vagina: a microscopic study," *Obstetrics and Gynecology* 12 (1958): 382–96; N. Martin-Alguacil, D. W. Pfaff, D. N. Shelly, and J. M. Schober, "Clitoral sexual arousal: an immunocytochemical and innervation study of the clitoris," *BJU International* 191 (2008): 1407–13; P. I. Vilimas, S.-Y. Yuan, R. V. Haberberger, and I. L. Gibbins, "Sensory innervation of the external genital tract of female guinea pigs and mice," *Journal of Sexual Medicine* 8 (2011): 1985–95; and Z. Halata and B. L. Munger, "The neuroanatomical basis for protopathic sensibility of the human glans penis," *Brain Research* 371 (1986): 205–30.

6. So what would it take to test the hypothesis that genital end bulbs have a special role in sexual touch sensation? You'd want a way to record their electrical signals to see if their firing correlates with sexual sensation. In addition, you'd need some way to selectively inactivate them: a drug, a genetically engineered virus, or a genetically engineered mouse. In a complementary fashion, you'd want a way to activate them without activating the other touch pathways in the same bit of skin. One way to do all of this is to try to find a gene that's selectively expressed in genital-end-bulb-bearing sensory neurons. Then, using genetic engineering, you could take the regulatory sequence from that gene (called the promoter) and use it to drive expression of proteins that activate, or suppress, neural signaling. You can even hook up this promoter to a protein that emits green light when a neuron is active. Then you just take a video image of the genital-end-bulb fibers to record their activity in various situations.

7. D. K. E. Van der Schoot and A. F. G. V. M. Ypma, "Seminal vesiculectomy to resolve defecation-induced orgasm," *BJU International* 90 (2002): 761–62.

8. B. R. Komisaruk, C. Gerdes, and B. Whipple, "'Complete' spinal cord injury does not block perceptual responses to genital self-stimulation in women," *Archives of Neurology* 54 (1997): 1513–20; and B. R. Komisaruk, B. Whipple, A. Crawford, A. Grimes, W.-C. Liu, A. Kalnin, and K. Mosier, "Brain activation during vaginocervical self-stimulation and orgasm in women

with complete spinal cord injury: fMRI evidence of mediation by the vagus nerves," *Brain Research* 1024 (2004): 77–88. It's worthwhile noting that the hypothesis that the vagus nerve conveys sexual touch information from the cervix and uterus to the brain is still a matter of debate among specialists.

9. N. Wolf, *Vagina: A Cultural History* (New York: HarperCollins, 2012), 26.

10. Wolf also believes that there is significantly greater individual variation in the sensory nerve wiring of the genital region in women when compared with men.

"This greater sexual neural complexity in women is because we have both reproductive and sexual parts, such as the cervix and uterus, that men don't have. There are many more neural networks extending from the female pelvis into the spinal cord than extend from the networks in the penis to the spinal cord." (*Vagina: A Cultural History*, 24.)

Yes, the sensory innervation of the penis alone is simpler than that of the entire female pelvis. If you compare the entire male pelvis to the entire female pelvis, there are important, obvious differences (no vagina or uterus to innervate in men and no testicles, scrotum, or prostate to innervate in women), but much less difference in the overall complexity of the sensory neural network. Most important, to my knowledge, there is no evidence that the *variation* in the fine structure of the sensory nerves is greater in either the genital or perigenital region of women. Many studies have shown that women do indeed have greater variation in response to sexual stimuli than men, but it remains unclear whether that is attributable in any part to variation in sensory innervation of the pelvis.

11. Wolf correctly states that "culture and upbringing" have an important role in one's individual sexual experiences and preferences and then goes on to contrast these influences with variation in "physical wiring." It should be noted that culture and upbringing and physical wiring are not entirely separate phenomena. It's not as if physical wiring derives solely from one's genome and culture and upbringing from life experience and the two never interact. As we've previously discussed, life experience, from early maternal attention to intensive practice of a musical instrument, can produce long-lasting changes in the fine structure and the cellular function of both the brain and the body. In short, nurture works through nature.

12. B. R. Komisaruk, N. Wise, E. Frangis, W.-C. Liu, K. Allen, and S. Brody, "Women's clitoris, vagina, and cervix mapped on the sensory cortex: fMRI evidence," *Journal of Sexual Medicine* 8 (2011): 2822–30.

13. Some researchers have found, in both men and women, that genital stimulation activates only the groin site on the body map, others have found activation only at the detached site beyond the toes, and some have found both. It's an ongoing argument in the scientific literature. The experiments designed to map the sensory representation of the genitals have used many different techniques. In some, stimulation was produced with a small vibrator or an electrical stimulator controlled by the experimenter. In others, self-stimulation with a dildo was used. Imaging of the brain was performed using varying techniques, from EEG, which has good temporal resolution but poor spatial resolution, to fMRI, which has moderate spatial resolution and poor temporal resolution, and others with their own sets of advantages and disadvantages. L. Michels, U. Mehnert, S. Boy, B. Schurch, and S.

Kollias, "The somatosensory representation of the human clitoris: an fMRI study," *NeuroImage* 49 (2010): 177–84; C. A. Kell, K. von Kriegstein, A. Rosler, A. Kleinschmidt, and H. Laufs, "The sensory cortical representation of the human penis: revisiting somatotopy in the male homunculus," *Journal of Neuroscience* 25 (2005): 5984–87; T. Allison, G. McCarthy, M. Luby, A. Puce, and D. D. Spencer, "Localization of function regions of human mesial cortex by somatosensory evoked potential recording and by cortical stimulation," *Electroencephalography and Clinical Neurophysiology* 100 (1996): 126–40; J. P. Makela, M. Illman, V. Jousmaki, J. Numminen, M. Lehecka, S. Salenius, N. Forss, and R. Hari, "Dorsal penile nerve stimulation elicits left-hemisphere dominant activation in the second somatosensory cortex," *Human Brain Mapping* 18 (2003): 90–99; and H. Nakagawa, T. Namima, M. Aizawa, K. Uichi, Y. Kaiho, K. Yoshikawa, S. Orikasa, and N. Nakasato, "Somatosensory-evoked magnetic fields elicited by dorsal penile, posterior tibial and median nerve stimulation," *Electroencephalography and Clinical Neurophysiology* 108 (1998): 57–61.

14. The observation that the foot, the nipple, and the genitals activate adjacent patches of primary somatosensory cortex has led some to suggest that cross talk between these regions might help explain the prevalence of foot-fetish sexual behavior. There's no evidence that speaks strongly either for or against this notion. However, it should be noted that seizures involving the foot part of the map, in the mesial surface of the postcentral gyrus, can produce sensations in the nipples, labia minora, and feet of women.

15. Indeed, the notion of penis-vagina intercourse without male orgasm has a long history in many cultures. It is called *coitus reservatus* in the European tradition and *cai yin pu yang* in the Taoist tradition. *Cai yin pu yang* means gathering and absorbing female essence (yin) by the male. Similarly, the practice of *karezza*, in which penis-vagina intercourse is controlled to bring both male and female partners near to orgasm, is promoted as a way of achieving mystical sexual ecstasy.

16. The idea of using a pleasure trajectory to compare sexual behavior with eating came from the following review article, which is worth a read if you'd like to dig into some technical details, particularly about how human sex compares to that of critters: J. R. Georgiadis, M. L. Kringelbach, and J. G. Pfaus, "Sex for fun: a synthesis of human and animal neurobiology," *Nature Reviews Urology* 9 (2012): 486–98.

17. There's a clear and compassionate FAQ about sex and spinal cord injury at http://www.sexsci.me/FAQ/.

18. For a more extended discussion of this topic, see my book *The Compass of Pleasure* (New York: Viking/Penguin, 2011), 105–11. For a complete meta-analysis of the scientific work on this topic, see M. L. Chivers, M. C. Seto, M. L. Lalumiere, E. Lann, and T. Grimbos, "Agreement of self-reported and genital measures of sexual arousal in men and women: a meta-analysis," *Archives of Sexual Behavior* 39 (2010): 5–56.

19. http://www.tampabay.com/features/humaninterest/persistent-genital-arousal-disorder-brings-woman-agony-not-ecstasy/1263980, story written by Leonora LaPeter Anton.

20. PGAD was originally called persistent sexual arousal syndrome (PSAS) until it was realized that it was a misnomer: It is not associated with any

increase in sexual arousal or desire. Another name for this condition is "restless genital syndrome," reinforcing the presumed link to restless leg syndrome. For a recent review of the literature on PGAD, see T. M. Facelle, H. Sadeghi-Nejad, and D. Goldmeir, "Persistent genital arousal disorder: characterization, etiology, and management," *Journal of Sexual Medicine* 10 (2013): 439–50.

21. J. Money, G. Wainwright, and D. Hingburger, *The Breathless Orgasm: A Lovemap Biography of Asphyxiophilia* (New York: Prometheus Books, 1991).

22. All this rapture and ecstasy means that orgasm has attracted the attention of religious authorities. On the one hand, orgasm has been seen as life-affirming and healthy. On the other hand, at least for men, orgasm can be viewed as debilitating, a loss of male energy and essence. Havelock Ellis, writing in 1910, reported the following recommended frequencies for male orgasm (female orgasm was not addressed): Hindu Authorities: three to six times per month; Martin Luther: twice per week; the Holy Koran: once per week; the Talmud: once per day to once per week depending upon one's occupation. H. Ellis, *Studies in the Psychology of Sex* (London: F. A. Davis, 1910).

23. For a comprehensive overview of this topic, see B. R. Komisaruk and B. Whipple, "Non-genital orgasms," *Sexual and Relationship Therapy* 26 (2011): 356–72. And for all of your many orgasm questions, I recommend this useful book: B. R. Komisaruk, B. Whipple, S. Naserzadeh, and C. Beyer-Flores, *The Orgasm Answer Guide* (Baltimore: Johns Hopkins University Press, 2010).

24. Some people can even have an orgasm from a hug while fully clothed. They are just wired that way.

25. J. Money, "Phantom orgasm in the dreams of paraplegic men and women," *Archives of General Psychiatry* 3 (1960): 373–82.

26. E. B. Vance and N. N. Wagner, "Written descriptions of orgasms: a study of sex differences," *Archives of Sexual Behavior* 5 (1976): 87–98.

27. S. K. Fistarol and P. H. Itin, "Diagnosis and treatment of lichen sclerosus: an update," *American Journal of Clinical Dermatology* 14 (2013): 27–47. Here is a useful informational Web site for people with lichen sclerosus: http://www.lichensclerosus.net/.

28. You can read Koedt's essay here: http://www.uic.edu/orgs/cwluherstory /CWLUArchive/vaginalmyth.html.

29. A. C. Kinsey, *Sexual Behavior in the Human Female* (New York: Pocket Books, 1953), 580.

30. Another structure on the anterior vaginal wall is the "G-spot," which has been proposed to play a special role in female sexual sensation. At present, the question of whether the G-spot is a well-defined anatomical entity is unresolved. For example, examination of tissue sections from cadavers has led to different conclusions by various researchers. If you'd like to read a recent review article exploring the controversy, see E. A. Jannini, B. Whipple, S. A. Kingsberg, O. Buisson, P. Foldès, and Y. Vardi, "Who's afraid of the G-spot?" *Journal of Sexual Medicine* 7 (2010): 25–34.

31. E. A. Jannini, A. Bubio-Casillas, B. Whipple, O. Buisson, B. Komisaruk, and S. Brody, "Female orgasm(s): one, two, several," *Journal of Sexual Medicine* 9 (2012): 956–65.

32. While we normally think of erection as being a prerequisite for orgasm, this is not entirely correct. For example, if the pudendal arteries become blocked, erection cannot occur, but stimulation of the penis can still produce orgasm in some cases.

33. M. Koeman, M. F. Van Driel, W. C. M. Weijmar Schultz, and H. J. A. Mensink, "Orgasm after radical prostatectomy," *British Journal of Urology* 77 (1996): 861–64.

34. And don't imagine that it's only gay or bisexual men who like stimulation of the anus, rectum, and prostate. My old pal C., who runs an Internet sex-toy shop, says, "You'll never go broke selling devices for straight guys to put in their butts."

35. G. Holstege, J. R. Georgiadis, A. M. J. Paans, L. C. Meiners, F. H. C. E. van der Graaf, and A. A. T. S. Reinders, "Brain activation during human male ejaculation," *Journal of Neuroscience* 23 (2003): 9185–93; J. R. Georgiadis and G. Holstege, "Human brain stimulation during sexual stimulation of the penis," *Journal of Comparative Neurology* 493 (2005): 33–38; and J. R. Georgiadis, R. Kortekaas, R. Kuipers, A. Nieuwenburg, J. Pruim, A. A. T. S. Reinders, and G. Holstege, "Regional cerebral blood flow changes associated with clitorally induced orgasm in healthy women," *European Journal of Neuroscience* 24 (2006): 3305–16.

36. The amygdala is best known for processing signals related to fear, but it also has roles in other types of emotional processing and learning.

37. J. Calleja, R. Carpizo, and J. Berciano, "Orgasmic epilepsy," *Epilepsia* 29 (1988): 635–39; Y. C. Chuang, T. K. Lin, C. C. Lui, S. D. Chen, and C. S. Chang, "Tooth-brushing epilepsy with ictal orgasms," *Seizure* 13 (2004): 179–82; and G. M. Remillard, F. Andermann, G. F. Testa, P. Gloor, M. Aube, J. B. Martin, W. Feindel, A. Guberman, and C. Simpson, "Sexual ictal manifestations predominate in women with temporal lobe epilepsy: a finding suggesting sexual dimorphism in the human brain," *Neurology* 33 (1983): 323–30.

CHAPTER FIVE: HOT CHILI PEPPERS, COOL MINT, AND VAMPIRE BATS

1. Chili peppers are the seedpods of five different cultivated species of the genus *Capsicum*. Archeological digs have indicated that people in present-day Mexico were eating chili peppers as early as six thousand years ago. Chili peppers were brought from the West Indies to Europe by Diego Álvarez Chanca, who was a doctor on Columbus's second voyage in 1493. Europeans, Africans, and Asians had black pepper in pre-Columbian times, but this is different from chili pepper. Black and white peppers come from the berries of the pepper plant *Piper nigrum*, which originated in south India. Black peppercorns are made by picking the pepper berries when they are partially ripe and then leaving them to dry. White peppercorns are picked when very ripe and subsequently soaked in brine to remove their dark outer shell, leaving only the white pepper seed. The main pungent compound from black pepper is piperene, which is chemically and functionally distinct from that of chili peppers. Some people in Asia during pre-Columbian times also made use of the seeds of the plants *Zanthoxylum simulans* and *Zanthoxylum bungeanum*, now known as Szechuan pepper. Szechuan peppercorns evoke a tingling, numbing sensation produced by its active

ingredient, alpha-sanshool. Like black pepper, Szechuan pepper is chemically and experientially distinct from chili pepper.

2. Perhaps people living in some remote locations in Borneo or Papua New Guinea have yet to experience chili peppers or hear them described as hot, but that's uncertain. There are some people living in the Amazon basin who have had little or no contact with the outside world, but chili peppers grow there so they may have encountered them locally.

3. Capsaicin is the main pungent compound in chili peppers, but there are also some other chemically related compounds called capsaicinoids. There are only five different species of *Capsicum* plants, but there are many different subspecies that have been created by humans, and these are known as cultivars. In much the same way that a poodle and a Labrador are both subspecies of the dog *Canis lupus* produced by human breeders, the wax pepper and the jalapeño pepper are both cultivars of the domesticated chili pepper *Capsicum annuum*.

4. The cold/menthol sensor, which we now call TRPM8, was first characterized by two different groups: D. D. McKemy, W. M. Neuhausser, and D. Julius, "Identification of a cold receptor reveals a general role for TRP channels in thermosensation," *Nature* 416 (2002): 52–58; and A. M. Peier, A. Moqrich, A. C. Hergarden, A. J. Reeve, D. A. Anderson, G. M. Story, T. J. Early, I. Dragoni, P. McIntyre, S. Bevan, and A. Patapoutian, "A TRP channel that senses cold and menthol," *Cell* 108 (2002): 705–15. And the heat/capsaicin receptor was first identified in this classic paper: M. J. Caterina, M. A. Schumacher, M. Tominaga, T. A. Roden, J. D. Levine, and D. Julius, "The capsaicin receptor: a heat-activated ion channel in the pain pathway," *Nature* 389 (1997): 816–24.

5. There are also synthetic compounds that can activate TRPM8 and produce a strong cooling sensation. One of these, originally called AG-3-5 but subsequently given the more evocative name of icilin, is about two-hundred-fold more potent than menthol. E. T. Wei and D. A. Seid, "AG-3-5: a chemical producing sensations of cold," *Journal of Pharmacy and Pharmacology* 35 (1983): 110–12.

6. P. Cesare and P. A. McNaughton, "A novel heat-activated current in nociceptive neurons and its sensitization by bradykinin," *Proceedings of the National Academy of Sciences of the USA* 93 (1996): 15435–39. The average shower temperature for adults in the United States is 107°F, just below the threshold for activation of TRPV1. There's some between-individual and within-individual variation, however. For a nice review of temperature-sensitive TRP channels, see L. Vay, C. Cu, and P. A. McNaughton, "The thermo-TRP ion channel family: properties and therapeutic implications," *British Journal of Pharmacology* 165 (2012): 787–801.

7. S. E. Jordt and D. Julius, "Molecular basis for species-specific sensitivity to 'hot' chili peppers," *Cell* 108 (2002): 421–30.

8. Birds that eat chili peppers have access to a food source that is shunned by almost all other animals. Among mammals, humans are the only species that intentionally consumes chili peppers, according to this useful review article: B. Nilius and G. Appendino, "Spices: the savory and beneficial science of pungency," *Review of Physiology, Biochemistry and Pharmacology* 164 (2013): 1–76.

9. The initial characterization of mice harboring a deletion of the TRPV1 gene was reported in these two papers: M. J. Caterina, A. Leffler, A. B. Malmberg, W. J. Martin, J. Trafton, K. R. Petersen-Zeitz, M. Koltzenberg, A. I. Basbaum, and D. Julius, "Impaired nociception and pain sensation in mice lacking the capsaicin receptor," *Science* 288 (2000): 306–13; and J. B. Davis, J. Grey, M. J. Gunthorpe, J. P. Hatcher, P. T. Davey, P. Overend, M. H. Harries, J. Latcham, C. Clapham, K. Atkinson, S. A. Hughes, K. Rance, E. Grau, A. J. Harper, P. L. Pugh, D. C. Rogers, S. Bingham, A. Randall, and S. A. Sheardown, "Vanilloid receptor-1 is essential for inflammatory thermal hyperalgesia," *Nature* 405 (2000): 183–87. Since then, the basic finding, that TRPV1 deletion abolishes capsaicin responses but only reduces heat responses and inflammation boosting of heat responses, has been replicated using specific drugs to block or desensitize TRPV1 and extended by recording the electrical signals from single sensory nerve fibers that innervate the skin.

10. S. M. Huang, X. Li, L. Yu, J. Wang, and M. J. Caterina, "TRPV3 and TRPV4 ion channels are not major contributors to mouse heat sensation," *Molecular Pain* 7 (2011): 37–47; and U. Park, N. Vastani, Y. Guan, S. N. Raja, M. Koltzenberg, and M. J. Caterina, "TRP vanilloid 2 knock-out mice are susceptible to perinatal lethality but display normal thermal and mechanical nociception," *Journal of Neuroscience* 31 (2011): 11425–36. Even when TRPV1 was inhibited, no further effects of TRPV2 or TRPV4+ TRPV3 deletion could be seen.

11. D. M. Bautista, J. Siemens, J. M. Glazer, P. R. Tsuruda, A. I. Basbaum, C. L. Stucky, S.-E. Jordt, and D. Julius, "The menthol receptor TRPM8 is the principal detector of environmental cold," *Nature* 448 (2007): 204–9; R. W. Colburn, M. L. Lubin, D. J. Stone Jr., Y. Wang, D. Lawrence, M. R. D'Andrea, M. R. Brandt, Y. Liu, C. M. Flores, and N. Qin, "Attenuated cold sensitivity in TRPM8 null mice," *Neuron* 54 (2007): 379–86; and A. Dhaka, A. N. Murray, J. Mathur, T. J. Early, M. J. Petrus, and A. Patapoutian, "TRPM8 is required for cold sensation in mice," *Neuron* 54 (2007): 371–78.

12. The present state of the scientific literature on the topic of TRPM8-independent sensors of cold is a bit of a mess. Some groups have claimed a role for the wasabi sensor gene TRPA1 (to be discussed soon), but others have failed to replicate these findings. To get a sense of the current state of the struggle, see D. McKemy, "The molecular and cellular basis of cold sensation," *ACS Chemical Neuroscience* 4 (2013): 238–47.

13. TRPA1 was originally called ANKTM1 before the nomenclature for TRP channels was standardized. S. E. Jordt, D. M. Bautista, H. H. Chuang, D. D. McKemy, P. M. Zygmunt, E. D. Högestätt, I. D. Meng, and D. Julius, "Mustard oils and cannabinoids excite sensory nerve fibers through the TRP channel ANKTM1," *Nature* 427 (2004): 260–65; and M. Bandell, G. M. Story, S. W. Hwang, V. Viswanath, S. R. Eid, M. J. Petrus, T. J. Early, and A. Patapoutian, "Noxious cold ion channel TRPA1 is activated by pungent compounds and bradykinin," *Neuron* 41 (2004): 849–57.

14. The chemicals allicin and DADS from garlic/onions are chemically similar to AITC from the mustard/horseradish/wasabi family of plants. D. M. Bautista, P. Mohaved, A. Hinman, H. E. Axelsson, O. Sterner, E. D. Högestätt, D. Julius, S.-E. Jordt, and P. M. Zygmunt, "Pungent products

from garlic activate the sensory ion channel TRPA1," *Proceedings of the National Academy of Sciences of the USA* 102 (2005): 12248–52.

15. Cooking onions not only degrades the TRPA1 activators allicin and DADS but also breaks down fructose polymers (fructans) into fructose monomers, yielding a sweet taste.

16. The best extra-virgin olive oils produce a localized pungency that's restricted to the throat. In fact, among experts, the best olive oils are called "two-cough" oils, as this particular effect is sought after. The compound in olive oil that produces this effect is a TRPA1-activaor called oleocanthol. TRPA1 is expressed in cells that line the pharynx, but not those of the mouth or tongue, and this distribution is suggested to give rise to the coughing effect of oleocanthol. C. Payrot de Gachons, K. Uchida, B. Bryant, A. Shima, J. B. Sperry, L. Dankulich-Nagrudny, M. Tominaga, A. B. Smith III, G. K. Beauchamp, and P. A. S. Bresline, "Unusual pungency from extra-virgin olive oil is attributable to restricted spatial expression of the receptor of oleocanthol," *Journal of Neuroscience* 31 (2011): 999–1099.

17. While two-pore potassium channels are widely expressed in both neural and nonneural cells, hyrdroxyl-alpha-sanshool excites only a particular subset of two-pore channels: those formed by the products of the genes KCNK3, KCNK9, and KCNK18. D. M. Bautista, Y. M. Sigal, A. D. Milstein, J. L. Garrison, J. A. Zorn, P. R. Tsuruda, R. A. Nicoll, and D. Julius, "Pungent agents from Szechuan peppers excite sensory neurons by inhibiting two-pore potassium channels," *Nature Neuroscience* 11 (2008): 772–79; and R. C. Lennertz, M. Tsunozaki, D. M. Bautista, and C. L. Stucky, "Physiological basis of tingling paresthesia evoked by hydroxyl-alpha-sanshool," *Journal of Neuroscience* 30 (2010): 4353–61. This paper claims that the sensation produced by Szechuan peppercorn extract on the lips resembles a 50-hertz vibration and is carried by Meissner fibers: N. Hagura, H. Barber, and P. Haggard, "Food vibrations: Asian spice sets lips trembling," *Proceedings of the Royal Society B: Biological Sciences* 280 (2013): 20131680.

18. Yes, there are cases of vampire bats preying upon sleeping humans. However, this is quite rare. All three species of vampire bat much prefer ungulates.

19. L. Kürten and U. Schmidt, "Thermoperception in the common vampire bat (*Desmodus rotundus*)," *Journal of Comparative Physiology* 146 (1982): 223–28.

20. E. O. Gracheva, J. F. Cordero-Morales, J. A. Gonzales-Carcacia, N. T. Ingolia, C. Manno, C. I. Aranguren, J. S. Weissman, and D. Julius, "Ganglion-specific splicing of TRPV1 underlies infrared sensation in vampire bats," *Nature* 476 (2011): 88–91. In the trigeminal ganglia of vampire bats, about 45 percent of the messenger RNA molecules directing translation of TRPV1 protein [were of the alternatively spliced short form], while this value was only about 5 percent in fruit bats. Not all of the neurons in the trigeminal ganglion of the vampire bat innervate the nasal pits, so the value of 45 percent short transcripts makes sense: We wouldn't expect it to be higher.

21. A nice summary of the early years of the work on snake infrared sensing can be found here: E. A. Newman and P. H. Hartline, "The infrared 'vision' of snakes," *Scientific American* 246 (1982): 116–27.

22. P. H. Hartline, L. Kass, and M. S. Loop, "Merging of modalities in the optic tectum: infrared and visual integration in rattlesnakes," *Science* 199 (1978): 1225–29; E. A. Newman and P. H. Hartline, "Integration of visual and infrared information in bimodal neurons of the rattlesnake optic tectum," *Science* 213 (1981): 789–91; and C. Moon, "Infrared-sensitive pit-organ and trigeminal ganglion in the crotaline snakes," *Anatomy & Cell Biology* (2011), doi:10.5115/acb.2011.44.1.8.

23. Both human and rattlesnake TRPA1 are sensitive to AITC, the pungent chemical from wasabi, but human TRPA1 is much more sensitive. There appears to have been a molecular trade-off between thermal and wasabi sensitivity. E. O. Gracheva, N. T. Ingolia, Y. M. Kelly, J. F. Cordero-Morales, G. Hollopeter, A. T. Chesler, E. E. Sánchez, J. C. Perez, J. S. Weissman, and D. Julius, "Molecular basis of infrared detection by snakes," *Nature* 464 (2010): 1006–11; and J. F. Cordero-Morales, E. O. Gracheva, and D. Julius, "Cytoplasmic ankyrin repeats of transient receptor potential A1 (TRPA1) dictate sensitivity to thermal and chemical stimuli," *Proceedings of the National Academy of Sciences of the USA* 108 (2011): 1184–91.

24. S. Yokoyama, A. Altun, and D. F. Denardo, "Molecular convergence of infrared vision in snakes," *Molecular Biology and Evolution* 28 (2011): 45–48; and J. Geng, D. Liang, K. Jiang, and P. Zhang, "Molecular evolution of the infrared sensory gene TRPA1 in snakes and implications for functional studies," *PLOS One* 6 (2011): e28644.

25. H. Schmitz and H. Bousack, "Modelling a historic oil-tank fire allows an estimate of the sensitivity of the infrared receptors in pyrophilous *Melanophila* beetles," *PLOS One* 7 (2012): e37627. There is a species of forest-fire-seeking beetle living in Australia (*Merimna atrata*) that also lays its eggs in freshly charred trees. The Australian fire beetle has four infrared sensors on each side of its body, toward the rear end. South American blood-sucking bugs called vinchuca (*Triatoma infestans*) are thought to sense infrared radiation from warm prey as well. A. L. Campbell, R. R. Naik, L. Sowards, and M. O. Stone, "Biological infrared imaging and sensing," *Micron* 33 (2002): 211–25; and H. Bleckmann, H. Schmitz, and G. von der Emde, "Nature as a model for technical sensors," *Journal of Comparative Physiology A* 190 (2004): 971–81.

26. B. R. Myers, Y. M. Sigal, and D. Julius, "Evolution of thermal response properties in a cold-activated TRP channel," *PLOS One* 4 (2009): e5741. It should be noted that TRPV1 and TRPM8 are not just present in the neurons that innervate the skin. They are also present in neurons that carry signals from deep tissues in the abdomen and so are likely to be important in sensing core temperature as well. In addition, TRPV1 (but not TRPM8) is present in the brain and may also have a role in the central processing of thermal information.

27. M. L. Loggia, M. C. Bushnell, M. Tétreault, I. Thiffault, C. Bhérer, N. K. Mohammed, A. A. Kuchinad, A. Laferrière, M.-J. Dicaire, L. Loisel, J. S. Mogil, and B. Brais, "Carriers of recessive WNK1/HSN2 mutations for hereditary sensory and autonomic neuropathy type 2 (HSAN2) are more sensitive to thermal stimuli," *Journal of Neuroscience* 29 (2009): 2162–66. There are more robust genetic links to pain perception, both thermal and other types of pain. We'll discuss these in the next chapter.

CHAPTER SIX: PAIN AND EMOTION

1. J. J. Cox, F. Reimann, A. K. Nicholas, G. Thornton, E. Roberts, K. Springell, G. Karbani, H. Jafri, J. Mannan, Y. Raashid, L. Al-Gazali, H. Mamamy, E. Valente, S. Gorman, R. Williams, D. P. McHale, J. N. Wood, F. M. Gribble, and C. G. Woods, "An *SCN9A* channelopathy causes congenital inability to experience pain," *Nature* 444 (2006): 894–98.
2. Ashlyn Blocker is a teenage girl living in Georgia who has congenital insensitivity to pain produced by a similar genetic mutation to that found in the British-Pakistani families. Her parents struggle to help her avoid injury. They have bought extra-thick carpet for the floors and furniture with rounded corners and edges. They struggle to balance their desire to protect her with the need for a teenager to have autonomy in her actions. Some people with congenital insensitivity to pain, like Ashlyn, have defects in sweating while others do not. The SCN9A gene is also expressed in the neurons that send messages to the skin to trigger sweating and flushing (vasodilation). The story of Ashlyn and her family was told by Justin Heckert in the *New York Times Magazine*: "The hazards of growing up painlessly," November 15, 2012.
3. C. R. Fertleman, C. D. Ferrie, J. Aicardi, N. A. F. Bednarek, O. Eeg-Olofsson, F. V. Elmslie, D. A. Griesemer, F. Goutières, M. Kirkpatrick, I. N. O. Malmros, M. Pollitzer, M. Rossiter, E. Roulet-Perez, R. Schubert, V. V. Smith, H. Testard, V. Wong, J. B. P. Stephenson, "Paroxysmal extreme pain disorder (previously familial rectal pain syndrome)," *Neurology* 69 (2007): 586–95; and R. Hayden and M. Grossman, "Rectal, ocular and submaxillary pain," *A.M.A. Journal of Diseases of Children* 97 (1959): 479–82. It's not clear why the jaw, anus, and eyes are the most common trigger zones for painful attacks in this disorder.
4. C. R. Fertleman, M. D. Baker, K. A. Parker, S. Moffat, F. V. Elmslie, B. Abrahamsen, J. Ostman, N. Klugbauer, J. N. Wood, R. M. Gardiner, and M. Rees, "*SCN9A* mutations in paroxysmal extreme pain disorder: allelic variants underlie distinct channel defects and phenotypes," *Neuron* 52 (2006): 767–74; and J.-S. Choi, F. Boralevi, O. Brissaud, J. Sánchez-Martín, R. H. M. Te Morsche, S. D. Dib-Hajj, J. P. H. Drenth, and S. G. Waxman, "Paroxysmal extreme pain disorder: a molecular lesion of peripheral neurons," *Nature Reviews Neurology* 7 (2011): 51–55. As in congenital insensitivity to pain, sufferers of paroxysmal extreme pain disorder have normal brain and nerve structure. It's the function that's changed. There's an important lesson here: Many diseases are not associated with obvious structural changes to organs or cells.
5. This calculation assumes that the A-delta and C-fiber conduction speeds in dinosaurs are approximately similar to those found in birds, the modern survivors of the dinosaur lineage.
6. Our understanding of the molecular sensors for pain is still woefully incomplete. There have been claims that TRPV2 and TRPA1 have a role in sensing heat and cold pain, respectively, but the results with mutant mice engineered to lack these sensors have been equivocal. Likewise, we don't know the molecular identity of the sensor for mechanical pain either, although there are some interesting candidates.

7. There is also a population of so-called "silent pain sensors." There are C-fiber neurons that are normally responsive to heat but become sensitive to mechanical stimulation only in the context of injury. There are also some C-fibers that are not polymodal and that can sense thermal pain. It's a complicated wiring diagram.

8. For an interesting reminiscence by the developer of the McGill Pain Questionnaire, see R. Melzack, "The McGill Pain Questionnaire," *Anesthesiology* 103 (2005): 199–202.

9. For more details of the layered structure of the spinal dorsal horn as it relates to pain perception, see A. L. Basbaum, D. M. Bautista, G. Scherrer, and D. Julius, "Cellular and molecular mechanisms of pain," *Cell* 139 (2009): 267–84.

10. How do we know that different types of pain information are kept largely separate in the initial stages of pain signaling? One example comes from experiments in mice in which TRPV1-expressing sensory neurons were selectively destroyed. These mice had a profound loss of heat pain but no change in their sensitivity to painful cold or mechanical stimuli. Conversely, selective destruction of a set of neurons expressing the gene MrgprD produces a deficit in mechanical pain with no change in heat sensitivity. D. J. Cavanaugh, H. Lee, L. Lo, S. D. Shields, M. J. Zylka, A. I. Basbaum, and D. J. Anderson, "Distinct subsets of unmyelinated primary sensory fibers mediate behavioral responses to noxious thermal and mechanical stimuli," *Proceedings of the National Academy of Sciences of the USA* 106 (2009): 9075–80.

11. Other pain pathways include one that activates the cerebellum, which is involved in fine-tuning of movements and subconscious predictive control of the body.

12. As you can see in figure 6.4, the anterior cingulate cortex and the insula can also be activated in two other ways—directly from the thalamus or from the secondary somatosensory cortex—so the spinomesencephalic pathway isn't the only way that these emotional pain centers get activated.

13. Most modes of brain scanning aren't fast enough to resolve the difference between first and second pain. However, the technique called magnetoencephalography is. In the first experiment to clearly resolve regional brain activation in first and second pain, a laser beam was directed to the back of the hand to provide a repeatable (but nonscarring) painful stimulus together with magnetoencephalography. M. Ploner, J. Gross, L. Timmermann, and A. Schnitzler, "Cortical representation of first and second pain sensation in humans," *Proceedings of the National Academy of Sciences of the USA* 99 (2001): 12444–48. For a more complete review of this topic, see P. Schweinhardt and M. C. Bushnell, "Pain imaging in health and disease—how far have we come?" *Journal of Clinical Investigation* 120 (2010): 3788–97.

14. The quotation is translated from the German in this original paper: P. Schilder and E. Stengel, "Schmerzasymbolie," *Zeitschrift fur die gesamte Neurologie und Psychiatrie* (1928): 113, 143–58. The translation comes from C. Klein, "What pain asymbolia really shows," http://tigger.uic.edu/~cvklein/papers/AsymboliaWebVers.pdf, 2001.

15. M. F. Seidel and N. E. Lane, "Control of arthritis pain with anti-nerve-growth factor: risk and benefit," *Current Rheumatology Reports* 14 (2012): 583–88. It may be that anti-NGF therapies will be useful, but only for a

subset of patients where benefit of pain relief outweighs the negative consequences of accelerated joint degeneration.

16. In addition to persistent increases in the strength of spinal cord pain synapses evoked by tissue damage, there is another set of changes in the spinal dorsal horn evoked by damage to sensory nerves. Both are discussed in this useful review: J. Sandkühler and D. Gruber-Schoffnegger, "Hyperalgesia by synaptic long-term potentiation (LTP): an update," *Current Opinion in Pharmacology* 12 (2012): 18–27.

17. These drugs are called N-methyl-D-aspartate (NMDA) receptor antagonists. They are one of several types of receptor for the excitatory neurotransmitter glutamate.

18. H. Flor, L. Nikolajsen, and T. Staehelin Jensen, "Phantom limb pain: a case of maladaptive CNS plasticity?" *Nature Reviews Neuroscience* 7 (2006): 873–81. It's not clear if all the changes in the somatosensory cortex associated with phantom limb pain are exclusively consequences of changes in the spinal cord or if some of them develop independently.

19. The interview from which this section was adapted was conducted by CNN anchor Catherine Callaway and aired on February 8, 2004. The interview was with both Private Turner and Sergeant Neil Mulvaney, who led the team that evacuated Turner and the other wounded soldiers while they remained under fire. The transcript may be found at http://transcripts.cnn .com/TRANSCRIPTS/0402/08/sm.09.html.

20. H. K. Beecher, "Pain in men wounded in battle," *Annals of Surgery* (1946): 96–105.

21. In this way, the brain reminds me of the G. W. Bush administration. If it hears about global warming, it sends down a message to NASA, "Shut up! I don't want to know about that." If it hears whispers of weapons of mass destruction in Iraq, it sends the message to the CIA, "Tell me more! Shout it out!"

22. If you're wondering how structures in the brain get named, the answer is that it's a hot mess. The early anatomists sometimes named structures using colors: locus coeruleus means "blue spot" (the same Latin root as cerulean blue pigment); or landmarks: periaqueductal gray is simply the gray matter "surrounding the aqueduct," which is a narrow channel carrying fluid that runs through the core of the brain stem.

23. The actions of rostroventral medulla and locus coeruleus neurons to modulate transmission of pain signals in the spinal cord are complex and our understanding is incomplete. Some descending connections are received on the presynaptic terminals of pain-transmitting peripheral neurons, others on their targets in the spinal dorsal horn, and yet others on interneurons, which themselves release enkaphalins. Locus coeruleus neurons release the transmitter norepinephrine and some rostroventral medulla neurons release serotonin. A portion of the pain-blunting action of serotonin- and norepinephrine-boosting drugs (SSRIs and SNRIs) appears to result from their action in the spinal dorsal horn. While a portion of morphine's analgesic action is in the periaqueductal gray region, local application of morphine to the spinal cord can also have useful analgesic effects, and so this is a common practice in hospitals, particularly after Cesarean births. For a useful overview of this topic, see M. H. Ossipov, G. O. Dussor, and F. Porreca,

"Central modulation of pain," *Journal of Clinical Investigation* 120 (2010): 3779–87.

24. Morphine and related drugs like oxycodone and heroin also activate the brain's pleasure circuit—in particular, a key region called the ventral tegmental area. This is the basis of the euphoric effect produced by opiates and, ultimately, the basis of its addictive qualities.

25. M. C. Bushnell, M. Čeko, and L. A. Low, "Cognitive and emotional control of pain and its disruption in chronic pain," *Nature Reviews Neuroscience* 14 (2013): 502–11.

26. There is now good evidence that depressed pain-free people are approximately two times more likely to develop chronic musculoskeletal pain than nondepressed pain-free people, and similar findings have emerged for anxiety disorders. Obviously, the causality can flow in both directions: Chronic pain can lead to persistent emotional disorders, too, creating a vicious cycle. D. D. Price, "Psychological and neural mechanisms of the affective dimension of pain," *Science* 288 (2000): 1769–72; and K. Wiech and I. Tracey, "The influence of negative emotions on pain: behavioral effects and neural mechanisms," *NeuroImage* 47 (2009): 987–94.

27. C. Helmchen, C. Mohr, C. Erdmann, F. Binkofski, and C. Büchel, "Neural activity related to self- versus externally generated painful stimuli reveals distinct differences in the lateral pain system," *Human Brain Mapping* 27 (2006): 755–65; and Y. Wang, J.-Y. Wang, and F. Luo, "Why self-induced pain feels less painful than externally generated pain: distinct brain activation patterns in self- and externally generated pain," *PLOS One* 8 (2011): e23536.

28. T. Koyama, J. G. McHaffie, P. J. Laurienti, and R. C. Coghill, "The subjective experience of pain: where expectations become reality," *Proceedings of the National Academy of Sciences* 102 (2005): 12950–55; J.-K. Zubieta and C. S. Stohler, "Neurobiological mechanisms of placebo responses," *Annals of the New York Academy of Sciences* 1156 (2009): 198–210; and M. Peciña, H. Azhar, T. M. Love, T. Lu, B. L. Fredrickson, C. S. Stohler, and J. K. Zubieta, "Personality trait predictors of placebo analgesia and neurobiological correlates," *Neuropsychopharmacology* 38 (2013): 639–46. There is also evidence suggesting that dopamine release in a region called the nucleus accumbens, a key portion of the pleasure circuit, is involved in placebo analgesia.

29. I. Tracey, "Getting the pain you expect: mechanisms of placebo, nocebo and reappraisal effects in humans," *Nature Medicine* 16 (2010): 1277–83; and S. Geuter and C. Büchel, "Facilitation of pain in the human spinal cord by nocebo treatment," *Journal of Neuroscience* 21 (2013): 13784–90. In a few early studies, the nocebo effect has been linked with activation of a region called the hippocampus, which is involved in memory for facts and events. Further work will determine if this is a causal relationship.

30. P. Posadzki, E. Ernst, R. Terry, and M. S. Lee, "Is yoga effective for pain? A systematic review of randomized clinical trials," *Complementary and Therapeutic Medicine* 19 (2011): 281–87; and A. Chiesa and A. Serretti, "Mindfulness-based interventions for chronic pain: a systematic review of the evidence," *Journal of Alternative and Complementary Medicine* 17 (2011): 83–93.

31. The more technical term for fight-or-flight response is activation of the sympathetic branch of the autonomic nervous system.

32. D. M. Perlman, T. V. Salomons, R. J. Davidson, and A. Lutz, "Differential effects on pain intensity and unpleasantness of two meditation practices," *Emotion* 10 (2010): 65–71; and A. Lutz, D. R. McFarlin, D. M. Perlman, T. V. Salomons, and R. J. Davidson, "Altered anterior insula activation during anticipation and experience of painful stimuli in expert meditators," *Neuro-Image* 64 (2013): 538–46. For you neuroanatomy mavens, the exact areas activated in these studies are the dorsal anterior insula and the anterior mid-cingulate cortex.

33. J. A. Grant, J. Courtemanche, and P. Rainville, "A non-elaborative mental stance and decoupling of executive and pain-related cortices predicts low pain sensitivity in Zen meditators," *Pain* 152 (2011): 150–56.

34. N. I. Eisenberger, "The pain of social disconnection: examining the shared neural underpinnings of physical and social pain," *Nature Reviews Neuroscience* 13 (2012): 421–34; and E. Kross, M. G. Berman, W. Mischel, E. E. Smith, and T. D. Wager, "Social rejection shares somatosensory representations with physical pain," *Proceedings of the National Academy of Sciences of the USA* 108 (2011): 6270–75. As a control, the recently dumped subjects were asked to look at a photo of a friend (of the same sex as their former partner) with whom they had never had a romantic or sexual relationship.

CHAPTER SEVEN: THE ITCHY AND SCRATCHY SHOW

1. The account of Semanza and Moses Katabarwa is adapted from an article by Elizabeth Landau that appeared on *CNN Health* on February 2, 2013, entitled "With River Blindness, You Never Sleep," www.cnn.com/2013/02 /02/health/river-blindness. The name of the bacterium that lives in the gut of the roundworm is called *Wolbachia*. It is a symbiont with the worm, meaning that neither the worm nor the bacterium harm each other. *Wolbachia* can be killed using certain antibiotics like doxycycline, so combined ivermectin and doxycycline therapy is sometimes used. The only hosts for mature *Onchocerca volvulus* worms are primates (mice, rats, and other typical lab species are not infected), and this has limited the study of river blindness in the laboratory.

2. In addition to poison ivy, urushiol is found in poison oak and poison sumac.

3. R. P. Tuckett, "Itch evoked by electrical stimulation of the skin," *Journal of Investigative Dermatology* 79 (1982): 368–73; and H. O. Handwerker, C. Forster, and C. Kirchhoff, "Discharge patterns of human C-fibers induced by itching and burning stimuli," *Journal of Neurophysiology* 66 (1991): 307–15.

4. There is reason to believe that coughing and itching are related sensations, both of which serve to remove an unwanted irritant. They may also share cellular and molecular circuitry in the nervous system. P. C. LaVinka and X. Dong, "Molecular signaling and targets from itch: lessons for cough," *Cough* 9 (2013): 8.

5. For a nice discussion of itch-signal processing theories, see A. Dhand and M. J. Aminoff, "The neurology of itch," *Brain*, advance online publication, 2013.

6. M. Schmelz, R. Schmidt, A. Bickel, H. O. Handwerker, and H. E. Torebjörk, "Specific C-receptors for itch in human skin," *Journal of Neuroscience* 17 (1997): 8003–8; and M. Schmelz, R. Schmidt, C. Weidner, M. Hilliges,

H. E. Torebjörk, and H. O. Handwerker, "Chemical response pattern of different classes of C-nociceptors to pruritogens and algogens," *Journal of Neurophysiology* 89 (2003): 2441–48. Itch appears to be conducted mostly by slow C-fibers. One of the ways we know that is by temporarily placing a ligature around the arm of a human subject. This will block the conduction of electrical signals through larger A-fibers but not small C-fibers and does not appear to significantly reduce most forms of itch sensation.

7. Q. Liu, Z. Tang, L. Surdenikova, S. Kim, K. N. Patel, A. Kim, F. Ru, Y. Guan, H.-J. Weng, Y. Geng, B. J. Undern, M. Kollarik, Z.-F. Chen, D. J. Anderson, and X. Dong, "Sensory neuron-specific GPCR Mrgprs are itch receptors mediating chloroquine-induced pruritus," *Cell* 139 (2009): 1353–65; and Q. Liu, H.-J. Weng, K. N. Patel, Z. Tang, H. Bai, M. Steinhoff, and X. Dong, "The distinct roles of two GPCRs, MrgprC11 and PAR2, in itch and hyperalgesia," *Science* 4 (2011): ra45.

8. The full name of NPPB is natriuretic peptide B. It's a peptide messenger that was initially identified as a regulator of cardiac function.

9. S. K. Mishra and M. A. Hoon, "The cells and circuitry for itch responses in mice," *Science* 340 (2013): 968–71.

10. L. Han, C. Ma, Q. Liu, H.-J. Weng, Y. Cui, Z. Tang, Y. Kim, H. Nie, L. Qu, K. N. Patel, Z. Li, B. McNeil, S. He, Y. Guan, B. Xiao, R. H. LaMotte, and X. Dong, "A subpopulation of nociceptors specifically linked to itch," *Nature Neuroscience* 16 (2013): 174–82.

11. Y.-G. Sun, Z.-Q. Zhao, X.-L. Meng, J. Yin, X. Y. Liu, and Z.-F. Chen, "Cellular basis of itch sensation," *Science* 325 (2009): 1531–34. This line of work has been a bit confusing. Initially, these authors believed that GRP was the neurotransmitter of the sensory cells in the dorsal root ganglion that were the primary itch receptors. However, more recent evidence has indicated that these cells do not express GRP, and instead GRP is a neurotransmitter for the neurons in the spinal cord that receive itch signals from NPPB-releasing cells.

12. S. R. Wilson, K. A. Gerhold, A. Bifolck-Fisher, Q. Liu, K. N. Patel, X. Dong, and D. M. Bautista, "TRPA1 is required for histamine-independent, Mas-related G protein-coupled receptor-mediated itch," *Nature Neuroscience* 14 (2011): 595–603.

13. Woody Guthrie wrote the lyrics to "Hesitating Beauty" and many other songs for which he never composed music. Years later, at the instigation of Woody's daughter Nora, Billy Bragg and Wilco were recruited to compose tunes and perform some of these abandoned Guthrie songs. "Hesitating Beauty" received a tune and a fine performance from Jeff Tweedy and his band Wilco. You can read the complete lyrics at http://www.metrolyrics.com/hesitating-beauty-lyrics-wilco.html.

14. G. A. Bin Saif, A. D. Papoiu, L. Banari, F. McGlone, S. G. Kwatra, Y. H. Chan, and G. Yosipovitch, "The pleasurability of scratching an itch: a psychophysical and topographical assessment," *British Journal of Dermatology*, EPUB ahead of print, 2012. In this experiment, the experimenter, not the subject, did the scratching in order to produce a constant intensity of scratching across subjects and skin areas. However, I can't help but wonder if the results would have been different if the subjects had been able to scratch themselves. In general, subjects will scratch themselves much more vigorously than an experimenter will.

15. In one study, monkeys received a small injection of histamine in the skin to evoke an itchy sensation. This caused activation of neurons in the spinothalamic tract of the spinal cord that convey itch signals to the brain. When the skin region including the histamine injection site was scratched by the experimenter, this reduced the activation of these spinal cord neurons by histamine. In control experiments, scratching did not reduce the electrical activation of neurons in the spinal cord activated by pain or light touch. S. Davidson, X. Zhang, S. G. Khasabov, D. A. Simone, and G. J. Giesler Jr., "Relief of itch by scratching: state-dependent inhibition of primate spinothalamic tract neurons," *Nature Neuroscience* 12 (2009): 544–46.

16. One interesting prediction of this evolutionary speculation is that those animals lacking the ability to scratch or otherwise dislodge insects or other parasites from the skin would possess fundamentally different itch-processing systems in the skin, spinal cord, and brain.

17. X.-Y. Liu, Z. C. Liu, Y.-G. Sun, M. Ross, S. Kim, F.-F. Tsai, Q.-F. Li, J. Jeffry, J.-Y. Kim, H. H. Loh, and Z.-F. Chen, "Unidirectional cross-activation of GPCR by MOR1D uncouples itch and analgesia induced by opioids," *Cell* 147 (2011): 447–58.

18. A. Drzezga, U. Darsow, R. D. Treede, H. Siebner, M. Frisch, F. Munz, F. Weilke, G. Ring, M. Schwaiger, and P. Bartenstein, "Central activation by histamine-induced itch: analogies to pain processing: a correlational analysis of O^{15} H_2O positron emission tomography studies," *Pain* 92 (2001): 295–305. More recent work has sought to dissociate the brain-response patterns to different types of itch: A. D. P. Papoiu, R. C. Coghill, R. A. Kraft, H. Wang, and G. Yosipovitch, "A tale of two itches. Common features and notable differences in brain activation induced by cowhage and histamine induced itch," *NeuroImage* 59 (2012): 3611–26.

19. A. L. Oaklander, "Common neuropathic itch syndromes," *Acta Dermato-Venereologica* 9 (2012): 118–25.

20. A. L. Oaklander, S. P. Cohen, and S. V. Y. Raju, "Intractable postherpetic itch and cutaneous deafferentation after facial shingles," *Pain* 96 (2002): 9–12. This case was also explored by Atul Gawande in a splendid piece that appeared in the *New Yorker* and which I have mined for some details: A. Gawande, "The itch," *New Yorker,* June 30, 2008, 58–65.

21. V. Niemeier, J. Kupfer, and U. Gieler, "Observations during an itch-inducing lecture," *Dermatology and Psychosomatics* 1 (2000): 15–18.

22. D. M. Lloyd, E. Hall, S. Hall, and F. McGlone, "Can itch-related visual stimuli alone provide a scratch response in healthy individuals?" *British Journal of Dermatology* 168 (2012): 106–111. It turns out that, like socially contagious yawning, contagious itch is not restricted to humans. Here's an interesting report of socially contagious itch in rhesus monkeys: A. N. Feneran, R. O'Donnell, A. Press, G. Yosipovitch, M. Cline, G. Dugan, A. D. P. Papoiu, L. A. Nattkemper, Y. H. Chan, and C. A. Shively, "Monkey see, monkey do: contagious itch in nonhuman primates," *Acta Dermato-Venereologica* 93 (2013): 27–29.

23. H. Holle, K. Warne, A. K. Seth, H. D. Critchley, and J. Ward, "Neural basis of contagious itch and why some people are more prone to it," *Proceedings of the National Academy of Sciences of the USA* 109 (2012): 19816–21.

24. André Gide, writing in his diary, March 19, 1931.

CHAPTER EIGHT: ILLUSION AND TRANSCENDENCE

1. M. J. Zigler, "An experimental study of the perception of clamminess," *American Journal of Psychology* 34 (1923): 550–61. Clamminess is an odd sensation. Its unpleasantness seems to derive in part from associations with the flesh of dead homeothermic animals or living poikilothermic (cold-blooded) ones.

2. Oil from seeds of the plant *Salvia hispanica*.

3. I. M. Bentley, "The synthetic experiment," *American Journal of Psychology* 11 (1900): 405–25. Bentley approached wetness perception using a very controlled situation in which the subject was not able to explore with his or her finger. In the real world, much of our perception of wetness using the hand involves active touch. More recently, experiments have been performed in which subjects are allowed to explore a wetted surface with the finger pad. In one of these, a particular signature of friction and acceleration (called stick slip) was found to be a reliable signature to distinguish water from thicker or greasy liquids. When stick-slip forces designed to mimic water were applied to the fingertip using a specially instrumented glass plate outfitted with ultrasonic stimulators, this could effectively simulate a water-coated surface. Subjects believed that the surface was wet even when it wasn't. Y. Nonomura, T. Miura, T. Miyashita, Y. Asao, H. Shirado, Y. Makino, and T. Maeno, "How to identify water from thickener aqueous solutions by touch," *Journal of the Royal Society Interface* 9 (1992): 1216–23.

4. There's a great experiment in which a similar tit-for-tat scenario unfolded in the lab. Two adult subjects faced each other, each resting the left index finger, palm up, in a molded depression. A metal bar on a hinge was then rested on top of each subject's finger. The hinge was outfitted with a sensor to measure the force delivered when the bar was pressed. Both subjects were given the same instructions: Exactly match the force of the tap on his finger that he receives with an equivalent tap when his turn comes. Neither subject knew the instructions given to the other. When the subjects took turns pressing on each other's fingers, the force applied always escalated dramatically, just like Natalie and Jacob in the bathroom door game. Each person said that he matched the force of the other's tap. When asked to guess the instructions given to the other person, each said, "You told the other person to press back twice as hard." The experiment was then modified such that the tap was produced by moving a joystick that controlled a motorized bar. The key difference between these two situations is that when the force is generated by bar pressing, making a stronger tap requires generating more force with the fingertip, which is sensed by the tapper. Instead, when the joystick is used, the motor does the work and there is no correlation between the force generated by the joystick-controlling finger and the force produced on the upturned finger of the other subject. In this configuration there was no significant force escalation. S. S. Shergill, P. M. Bays, C. D. Frith, and D. M. Wolpert, "Two eyes for an eye: the neuroscience of force escalation," *Science* 301 (2003): 187. This study is discussed in the context of cerebellar function in a previous book of mine: D. J. Linden, *The Accidental Mind* (Cambridge, MA: Harvard/Belknap Press, 1997),9–13. In these scenarios, estimation of force is not entirely dependent upon touch signals. Sensors in the joints and muscles also contribute.

5. Attenuation of self-generated touch is just one example of a more general phenomenon in the brain to reduce or ignore the sensory consequences of self-generated motion. When we move our vocal cords to speak, this temporarily reduces our ability to perceive sound, particularly other voices: It's difficult to speak and listen at the same time. In the visual system there's an even more dramatic example. Your eyes continually dart about using tiny fast motions called saccades. If we yanked a video camera around to mimic the way your eyes move and then played the resultant movie on a screen, the jerky image would be very difficult to watch and interpret. Instead, your brain has a circuit that totally blanks out your conscious perception of the stream of visual information during these saccades and fuses together the more stable parts of the visual information stream to create a useful visual scene that appears seamless and coherent. Movement in the visual world that results from rapid eye movements is not just attenuated; it's totally suppressed.

6. For a nerd like me, writing the phrase "computerized mechanical tickler" is deeply satisfying.

7. S.-J. Blakemore, D. M. Wolpert, and C. D. Frith, "Central cancellation of self-produced tickle sensation," *Nature Neuroscience* 1 (1998): 635–40; S.-J. Blakemore, D. M. Wolpert, and C. D. Frith, "The cerebellum contributes to somatosensory cortical activity during self-produced tactile stimulation," *NeuroImage* 10 (1999): 448–59; and S.-J. Blakemore, D. M. Wolpert, and C. D. Frith, "Why can't you tickle yourself?" *Neuroreport* 11 (2000): 11–16. Are you thinking what I'm thinking? If one can make a device that can make self-tickling feel ticklish by introducing delays and kinetic changes between manual stimulation and one's own skin, then what would happen if you applied this strategy to enhance the function of sex toys? Imagine it: The cerebellar-attenuating touch-enhancing vibrator. I'm going on Kickstarter now. Back in a few.

8. Just like other touch sensations we've discussed, tickle can be influenced by cognitive and emotional factors. It's hard to effectively tickle someone who is very sad or extremely angry. I remember playing a game as a teenager in which the goal was not to laugh when tickled. In my social group, T. was the master of this game and I was not particularly good, laughing easily. In an attempt to improve, I studied his face during the tickle-resistance game and saw that he had worked himself up into a terrible scowl. I tried it out and it worked: If you could make yourself feel angry using techniques of method acting (recalling some past injustice is effective), then you could render yourself immune to tickle.

9. M. Blagrove, S.-J. Blakemore, and B. R. J. Thayer, "The ability to self-tickle following rapid eye movement sleep dreaming," *Consciousness and Cognition* 15 (2006): 285–94.

10. The muscles controlled by your spinal cord become totally limp during REM sleep, but those in your head do not as they are controlled by centers in the brain stem. That is what gives REM sleep its name: REM = rapid eye movements.

11. B. A. Sharpless and J. P. Barber, "Lifetime prevalence rates of sleep paralysis: a systematic review," *Sleep Medicine Reviews* 15 (2011): 311–15. These authors estimate that 8 percent of the general population has had at least one

episode of sleep paralysis. This rises to 28 percent of students and 32 percent of psychiatric patients.

12. If any of you hipsters out there want to call your alt-country band Cutaneous Rabbit, I wouldn't object. The illusion was first described and named in this scientific paper: F. A. Geldard and C. E. Sherrick, "The cutaneous 'rabbit': a perceptual illusion," *Science* 178 (1972): 178–79. If you enjoy tactile illusions, there's a nice feature story on them that appeared in the magazine *New Scientist* on March 11, 2009, written by Graham Lawton: http://www .newscientist.com/special/tactile-illusions. For the real tactile illusion mavens among you, this review provides a rather complete survey: S. J. Lederman and L. A. Jones, "Tactile and haptic illusions," *IEEE Transactions on Haptics* 4 (2011): 273–94.

13. Various scientists have now spent a lot of time probing the details of the cutaneous rabbit illusion. For example, the illusion holds even if the body sites are not contiguous: When taps are given first to the ring finger and then to the index finger, the perception will be that the stimulus hops, landing on the middle finger on the way. J. P. Warren, M. Santello, and S. I. Helms Tillery, "Electrotactile stimuli delivered across fingertips including the cutaneous rabbit effect," *Experimental Brain Research* 206 (2010): 419–26. In another study, when the subjects held a stick such that it was laid across the tips of their index fingers and received the taps via the stick, they reported sensing the illusory taps in the space between the actual stimulus locations—that is, along the stick. This suggests that the cutaneous rabbit illusion involves not only the brain's map of the body but also the brain's representation of the extended body schema that results from body-object interactions. That is so cool. M. Miyazaki, M. Hirashima, and D. Zozaki, "The 'cutaneous rabbit' hopping out of the body," *Journal of Neuroscience* 30 (2010): 1856–60.

14. D. Goldreich, "A Bayesian perceptual model replicates the cutaneous rabbit and other tactile spatiotemporal illusions," *PLOS One* 2 (2007): e333; and D. Goldreich and J. Tong, "Prediction, postdiction, and perceptual length contraction: a Bayesian low-speed prior captures the cutaneous rabbit and related illusions," *Frontiers in Psychology* 4 (2013): 221.

15. V. Jousmäki and R. Hari, "Parchment-skin illusion: sound-biased touch," *Current Biology* 8 (1998): R190; and S. Guest, C. Catmur, D. Lloyd, and C. Spence, "Audiotactile interactions in roughness perception," *Experimental Brain Research* 146 (2002): 161–71.

16. E. H. Weber, in R. Wagner (ed.), *Handwörterbuch der Physiologie* 3 (1846): 481–588; and J. C. Stevens and B. G. Green, "Temperature-touch interaction: Weber's phenomenon revisited," *Sensory Processes* 2 (1978): 206–19. It's interesting that the weight-temperature illusion works only for passive perception of weight. It doesn't work for a coin that you heft in your palm, for example.

17. K. O. Johnson, I. Darian-Smith, and C. LaMotte, "Peripheral neural determinants in man: a correlative study of responses to cooling skin," *Journal of Neurophysiology* 36 (1973): 347–70.

18. Or maybe it is for you—different strokes and all that.

19. All of the component parts of this narrative are well understood. That said, we don't yet have good dual brain-scanning studies of interpersonal touch.

20. Sandra and Matthew Blakeslee do an unusually good job of discussing the plasticity of body schema in their book *The Body Has a Mind of Its Own* (New York: Random House, 2007).
21. M. B. Rothberg, A. Arora, J. Hermann, R. Kleppel, P. St. Marie, and P. Visitainer, "Phantom vibration syndrome among the medical staff: a cross sectional survey," *BMJ* 341 (2010): c6914. Similar results were found among medical interns in Taiwan: Y.-H. Lin, S.-H. Lin, P. Li, W.-L. Huang, and C.-Y. Chen, "Prevalent hallucinations during medical internships: phantom vibration and ringing syndromes," *PLOS One* 8 (2013): e65152.
22. If Shakespeare were working today and he was considering the issue of mysterious touch sensations, he might have written this instead:

    I feel a buzzing on my phone
    Though 'twas left at yonder home

23. R. N. Jamison, K. O. Anderson, and M. A. Slater, "Weather changes and pain: perceived influence of local climate on pain complaint in chronic pain patients," *Pain* 61 (1995): 309–15.
24. D. A. Redelmeier and A. Tversky, "On the belief that arthritis pain is related to the weather," *Proceedings of the National Academy of Sciences of the USA* 93 (1996): 2895–96.

INDEX